LONG-TERM SURVIVAL IN THE COMING

DARK AGE

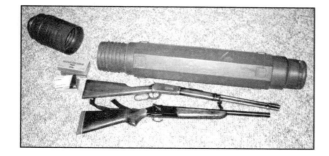

LONG-TERM SURVIVAL IN THE COMING

DARK AGE

Preparing to Live after Society Crumbles

James Ballou
foreword by Ragnar Benson

Paladin Press • Boulder, Colorado

Also by James Ballou:
Makeshift Workshop Skills for Survival and Self-Reliance
More Makeshift Workshop Skills for Survival and Self-Reliance

Long-Term Survival in the Coming Dark Age:
Preparing to Live after Society Crumbles
by James Ballou

Copyright © 2007 by James Ballou

ISBN 13: 978-1-58160-575-4
Printed in the United States of America

Published by Paladin Press, a division of
Paladin Enterprises, Inc.
Gunbarrel Tech Center
7077 Winchester Circle
Boulder, Colorado 80301 USA
+1.303.443.7250

Direct inquiries and/or orders to the above address.

PALADIN, PALADIN PRESS, and the "horse head" design
are trademarks belonging to Paladin Enterprises and
registered in United States Patent and Trademark Office.

Visit our website at www.paladin-press.com

To my dad, Gene Ballou, who gave me lots of
help and encouragement while I wrote this book.

Table of Contents

Introduction	Considering the Coming Dark Age	*1*
Chapter 1	Basic Dark-Age Preparations	*3*
Chapter 2	Caching Supplies	*17*
Chapter 3	The Survival Workshop	*29*
Chapter 4	Recycle and Salvage Everything	*65*
Chapter 5	Making Fire	*75*
Chapter 6	Making and Using Cordage	*83*
Chapter 7	Making Clothing	*93*
Chapter 8	A Barter Economy	*109*
Chapter 9	Adapting to a New Social Order	*117*

Foreword

Survival, more currently called "preparedness," carries with it great personal responsibility.

So daunting is the thought of this much personal responsibility that many people simply refuse to even consider the realities and implications of a survival economy. This is especially true when forced to contemplate long term or extended life in survival mode as is proposed by author James Ballou.

Europeans tend (perhaps erroneously) to equate survival entirely with eating berries and twigs while regressing to primitive, labor-intense means of production. Americans, in sharp contrast, view survival as a means of getting along nicely without government intervention, control, or help—if, in reality, there is such a thing.

History is, of course, on the side of the Americans, who note that just in the last century we have examples of Berlin, Madrid, Jerusalem, Stalingrad, Warsaw, Nanking, Tokyo, Beirut, Manila, and many other cities wherein a total collapse occurred. Citizens were summarily left to their own devices for an extended period of time. Many, so circumstanced, were unable to take personal responsibility for their own well being and simply disappeared.

Reasons for, and causes of, a collapse are many and varied. Wisely, Ballou does not overly entangle himself with these, or the likelihood of their occurring. Suffice it to say that these are dangerous times. The likelihood of some sort of national or even international catastrophe with which our government is incapable of coping seems more and more probable.

Ballou is concerned that wise people prepare for the very long term. The cost of being prepared need not be excessive; the cost of being unprepared can be huge.

Critics may point out that most of my examples of cities in crisis were not for the long term. Yet, three years or more is a very, very long term for those with absolutely no preparation in terms of skills, plans, reference library, or goods.

And what about the modern-day example of Cuba? There the economy has been in a state of collapse for almost 50 years. A very few innovative, creative, hard-working Cubans are still making it, but only because they have adapted many or most of the plans and devices suggested by Ballou.

A cornerstone of preparedness plans must include the absolute truth that at least one source of life's vital goods must be renewable.

No matter how many goods—especially hardware—remain to be scrounged, some items must be made new. In that regard, we are going to have to create, build, and improve, frequently creating vital survival goods out of what seems to others to be thin air.

Long-Term Survival in the Coming Dark Age very nicely addresses many of the skills and talents that many of us will need to acquire if we are to be among the elite group known as "survivors."

Specialization, as practiced by North American Indians as well as early European settlers, trading, scrounging, and making do with substitute goods will all be of vital importance. Even if readers are unwilling, at this time, to expend the time and effort to master many of the skills outlined in this book, they will—at least—be sensitized to the need for such should a Dark Ages scenario pop up on the horizon.

"Better late than never" may not always be much better, but it is still something.

Having information such as this in one's library is always of great value, even if no immediate action is taken. Yet, acting now on this information—while there is still time—is of much greater value, but only for those sufficiently brave and independent to say, "I will make it through, completely on my own."

—Ragnar Benson
2006

Considering the Coming Dark Age

Dark Ages: The period between about 500 and 1000, which was marked by frequent warfare and a virtual disappearance of urban life.[1]

At the time of this writing, it is probably safe to say that most Americans believe with confidence that civilization and technology will continue to advance at a progressively accelerating pace. Our history over the past 200 or more years would seem to indicate that this is what we can expect.

But the future can certainly be unpredictable. Before September 2001, who would ever have dreamed that commercial passenger planes would smash into and destroy the World Trade Center buildings? Before December 2004, who could have imagined the tsunami that killed well over 200,000 people along the coastlines of Indonesia and several South Asian countries?

When we consider some of the events of the 20th century—two world wars and numerous other bloody conflicts, the Great Depression of the 1930s, the AIDS epidemic, and the series of natural disasters just since 1989—it does not seem unreasonable to expect that some major world-changing events await us in the 21st century. More countries have nuclear weapon capabilities now than ever before, and the Earth's climate seems to be changing in some very strange ways.

Civilized societies function on a system of interdependence: A government depends on the taxes collected from its citizens; citizens, in turn, depend on their jobs for the income to pay those taxes and to buy goods and services essential to their survival; their employers depend on customers who purchase their products and services. Most citizens depend on motor transportation to get to work, which means depending on vehicle manufacturers, fueling stations, repair shops, and roadways, which are all paid for with money from the people who drive the cars, buses, and trucks that move the people and goods to work, to market, or out shopping to buy the things that create the jobs that pay the taxes to the government. And on and on.

So really, every part of a society depends on the other parts of the society for its existence, either directly or indirectly. The whole system works like a complex machine, and if any major component of the machine goes down for any reason, it could have a disabling effect upon the entire works.

Just imagine, for example, what would happen if a really nasty computer virus—one that was capable of circumventing the various virus protection programs—started spreading rapidly and, say, roughly 80 percent of all computers, including business and banking computers, personal computers, and government computers crashed all at the same time. What if a nuclear bomb detonated in New York City, destroying hundreds of buildings, killing people by the thousands or possibly millions, vaporizing Wall Street, and making the collapse of the Twin Towers look like a pinprick by comparison? Could you imagine the chaos that would follow?

The possible catastrophic scenarios may be endless, but the point is that modern civilization has certain vulnerabilities and is perhaps more fragile than we generally realize. If the New York Stock Exchange suddenly disappeared, what would that do to our national economy? If our economy collapsed, what effect would it have on the rest of the world? If a deadly tsunami were to devastate the eastern seaboard of the United States, how would it affect the rest of the country? The few scenarios mentioned here could even be mild compared to the perils our world may encounter in the years to come.

This book is intended to help readers prepare themselves for life without the type of social order we often take for granted. Organized cities, municipalities, governments, power stations, factories, stores, computers, and the general infrastructure of civilization may not always exist the way we know these things today. An array of challenges likely encountered by survivors in the months and years following a cataclysmic event will be discussed in the chapters of this book, with the end goal of finding practical solutions to some possibly exceptional problems.

ENDNOTE

1. "Dark Ages." Encyclopedia Britannica Premium Service. http://www.britannica.com/eb/article-9028782. (Accessed July 31, 2006.)

CHAPTER 1

Basic Dark-Age Preparations

Mountains and forests might offer survivors the best chances for survival, with wild game, fresh air, fresh water, lumber, and a place to hide when the turmoil begins. The cities may become violent slum areas for the most part, plagued with diseases.

The best time to prepare for hard times is usually during good times, for obvious reasons. But relatively few people are forward-thinking enough to prepare for a calamity that might not even happen. Some use the excuse, "Remember all the doom-and-gloom predictions about Y2K?", dismissing as fantasy the possibility of other problems. Consequently, plenty of excellent opportunities for getting ready are often missed, and every year the usual wildfires, floods, hurricanes, and other natural disasters catch people completely unprepared. Our world has a way of making unprepared people suffer.

According to a *TIME* magazine article published on the one-year anniversary of hurricane Katrina, 91 percent of Americans live in areas with a moderate-to-high risk of danger from earthquakes, volcanoes, tornadoes, wildfires, hurricanes, flooding, high-wind damage, or terrorism. But instead of preparing for these potential disasters, they tend to ignore them. "There are four stages of denial," says Eric Holdeman, director of emergency

3

management for Seattle's King County, which faces a significant earthquake threat. "One is, it won't happen. Two is, if it does happen, it won't happen to me. Three: if it does happen to me, it won't be that bad. And four: if it happens to me and it's bad, there's nothing I can do to stop it anyway."[1]

WHY PREPARE?

The people in this fourth group often seem bewildered by those who make a survival plan; they seem to believe only paranoid doomsday pessimists plan for a perilous future.

In my view, however, the best preparedness strategy isn't organized by a pessimist but by an optimist. An optimist will take action now in order to make his future better; he believes he *can* survive and is willing to invest some effort toward ensuring the best possible living conditions for himself and his family in spite of anything that may happen around him. If you weigh the disadvantages of being unprepared against the disadvantages (if any) of making yourself even a little bit prepared, which way do you think the scales will tip?

If we are certain something is going to happen, most of us prepare for it the best way that we can—parents of newborns start college funds, workers put their money into 401(k) accounts or other retirement plans—but when we can only speculate on the nature of a coming event, it is more difficult to prepare for it. In a sense we are all like ships sailing out into the ocean at night without navigation or observation equipment, and none of us knows where the icebergs are drifting. Each of us is a potential Titanic. This book is intended to help you avoid sinking into the abyss.

WHAT'S OUR STRATEGY?

Preparing for a future Dark Age will be quite different from preparing for a bad storm or any other local or short-term predicament. An "age" implies a period of some years. Here we will focus primarily on the long-term survival issues under the worst possible conditions.

You can develop a strategy for making yourself better prepared than most people will be at shock and panic time. If your strategy is sound, it should help you through the first phase at least. After that, survivors will have to adapt to a very different kind of world than we know today.

What Could Happen?

If you live in a city, try to envision your neighborhood without the weekly garbage pickup, reliable electricity, or a functioning sewer system. Running water may be scarce, and responding fire and police protection could be very difficult to manage without the resources of an organized government, which may ultimately be replaced by local gangs. The remaining elements of human society may eventually regroup into a tribal system, reminiscent of history's more primitive cultures.

Normal commerce may also be so severely disrupted by a major catastrophe that supermarkets and shopping malls are no longer in operation. These businesses rely heavily on electricity, computers, security, a labor force, and shipping, and any or all of these things may be unreliable at best, and nonexistent in the worst of situations. Imagine not being able to buy fresh produce, or toilet paper, or milk, at the store where you always used to shop. In the first weeks following a catastrophe, store shelves may be picked clean by desperate looters. Under less-than-ideal circumstances it could take a long time to restore a lost system.

If the government of the United States suddenly collapsed, the American dollar would only be as valuable as any other green-colored strip of paper. It would have essentially no buying power at all. The same is true for the currency of any other country. Unlike the currency of years past, which was minted from inherently valuable precious metals, the only thing that bestows any value at all upon modern money is the faith people have in their government. Since commerce works best when a standard medium of exchange can be relied upon, traders without a stable currency will be forced to barter with one another. The barter system might work well between individuals, but it certainly is not industry-friendly. Banks, manufacturers, retailers, restaurants, and other businesses would struggle severely in a barter economy.

A good preparedness strategy will take all of these matters into consideration. Basic living would demand different occupational skills than most of us presently apply in our normal routines. Where now we concern ourselves with maintaining our automobiles, paying our monthly bills, getting to work or school on time, budgeting our incomes, and filing our taxes, in a harsher environment we will have other priorities. We might have to stalk, kill, and butcher any meat that we eat ourselves because the grocery store is closed. We might have to move about on foot or on a bicycle because the gas stations are closed, or maybe the fuel is unaffordable for most of us. And if we have anything of utility or survival value, we might have to protect it by force, because the police may no longer serve the people.

Short Term vs. Long Term

Preparing for the short term should be fairly easy. It will involve stockpiling certain essentials such as storable foods, bottled water, extra cash, blankets, batteries, fuel and fuel preservatives, and medicines. Basic emergency and first-aid skills would be valuable. Keeping bags packed for travel might be handy in the event of a rapid evacuation. Things like gas masks, fire extinguishers, road flares, inflatable rafts, tire pumps, winter clothing, ropes, tarps, water filters, and other miscellaneous emergency survival gear could be useful immediately following a disaster of any proportion.

Long-term survival presents additional issues. Eventually, stockpiled resources will dwindle. Water supplies may dry up or become contaminated. Motor vehicles, machines, appliances, and other manufac-

tured products will eventually wear out. The progressive deterioration of sanitation would be likely. For how long such conditions could last is anybody's guess. Rebuilding the infrastructure and restoring technology would most likely become civilization's main priorities, but day-to-day survival needs will take precedence for some time.

The natural tendency of social order is for the strongest to dominate and—at least for a while—those with the greatest firepower or superior weapons could become the rulers in the barbaric new society. (The issues of individual security and survival weaponry in times of anarchy are very important topics; I hope to cover them extensively in another book.)

Where Do I Start?

Any preparations you begin making now should be made in anticipation of a world with different demands. Obviously, anyone with the financial resources and initiative to invest now in rural acreage is probably wise to do so. You could become quite self-sufficient with a private well, septic system, solar or other alternative power source, backup generators, wood stove and several years' supply of firewood, fruit cellar, fruit trees, greenhouse, chickens and cows, several late-model 4x4 utility pickups, fuel storage tanks, tool shop, a dozen hunting and military weapons with thousands of rounds of ammunition, and a basement full of storable foods and other groceries. But while most of us will have to make do with less, there are still many things we can do to get through this difficult time.

People who survive the initial stages of a severe global crisis will have to be resourceful to continue surviving. A certain degree of ingenuity will evolve by necessity where it didn't exist before, we can be sure. The ideas and concepts outlined in this book should be helpful to almost anyone who studies and remembers them in the coming years, should our modern world unravel the way I believe it so easily could.

While you would be concerned quite a lot with matters of self-defense in a time after a global cataclysm, no doubt, and food and supplies might be scarce and precious, you will also need practical skills and a useful education. Pursuits generally popular in a time of prosperity, such as performing arts, may no longer be in such high demand. You might find yourself in desperate need of a doctor or dentist who can help you with whatever improvised tools he might scrounge up. A mechanic who is able to keep a generator or a vehicle running after it's considered worn out could be a very important person all of a sudden; people with practical skills will always be valuable.

You could start educating yourself right away in as many subjects as might be useful to your future survival in a harsh environment. Subjects related to technology and science, mechanics, electronics, medicine, agriculture, or engineering might be especially useful, as well as wilderness survival skills, self-defense, and warfare. If knowledge is power, then you have access to unlimited power right now. You might as well educate yourself while the vast educational resources are still available to you.

THE BASICS

In this chapter dealing with basic, long-term survival preparations, a somewhat comprehensive discussion on equipment might be useful. A list of provisions will be presented here, with some of the gear to be reviewed further in subsequent chapters.

As already noted, it's fairly easy to lay in enough supplies to get you through a couple of days or even weeks after a disaster. Equipping for an unspecified period of years presents a greater challenge. Also, what you read here represents one man's ideas about basic long-term survival gear. The reader should be mindful that one person's (or family's) situation, geographic location, shopping budget, storage or load-carrying capacity, and medical or dietary requirements may differ substantially from another's.

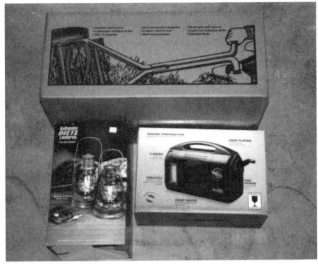

Low-tech devices might be the most practical in a Dark Age.

Vegetable seeds weigh next to nothing, take up little space, and are cheap now compared to their potential value after a collapse.

An assortment of old steel traps might serve as barter or as food collectors during a Dark Age.

Duffel bags provide a convenient and portable way to store gear.

Survival belt packs can be used for day hunts and scouting forays.

Small caches of food and groceries could relieve some of the initial stress of a catastrophe.

Containing Your Supplies

I begin my list with a way to contain the provisions. Some may prefer to stock the shelves of a pantry or laundry room with a hoard of supplies, but I like the mobility advantage of keeping a big bag, or maybe several big bags, packed and ready to roll. The dilemma here is that while you'd probably like a room full of supplies to last even the first year, on the other hand, if you needed to pack up and evacuate in a hurry, it would be awfully difficult to clear a bunch of shelves and then quickly load the loose provisions into your vehicle. You can save yourself a lot of stress in a crisis and ensure that you won't forget the basics you want by packing ahead of time.

I advise using large duffel bags constructed of tough canvas or high-denier nylon. Backpacks are also very good, although most lack the capacity for the amount of gear we are talking about. Hard-sided suitcases are easy to handle, rugged, and offer good protection for your gear, but they just don't have enough room inside for our purposes. A giant duffel bag the size of a bathtub will hold a lot of gear and still be portable.

There are also certain advantages to having fewer supplies, such as increased mobility and less to worry about trying to hide or protect. In some ways, you might feel freer with fewer possessions. Being able to pick up and move quickly and quietly at any time could be an important aspect to your survival, depending on your situation. With this perspective in mind, fanny packs and small backpacks might also be part of your kit for short-range travels and forays.

Another option is to keep caches of supplies stashed away, hidden or buried in locations you have confidence in being able to access after a disaster. This would certainly solve most of your portability problems. However, the use of supply caches involves a whole new set of issues to consider. More attention will be given to the various aspects of caching in the next chapter.

Any food you choose to include in your long-term kit should have good nutritional

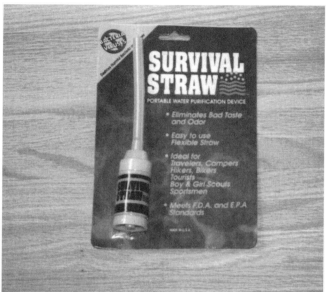

Various water filter devices are sold at most outdoor supply stores.

value, be well sealed in its packaging, have long shelf life, and be as lightweight and compact as possible. Dried beans, freeze-dried fruit, military rations/MREs, certain grains, and vitamin tablets are all popular survival foods.

Water
Drinkable water could be an especially precious commodity after a major disaster. The obvious problem here is that you will soon need more than you could ever hope to carry. Nevertheless, I believe a kit is incomplete without at least several quarts of pure bottled water to hydrate your body in the initial stages of an event. Some type of canteen, collapsible water bladder, or other water-carrying system might be particularly useful.

Perhaps even more valuable than your bottled water in the long run would be a water filtering/purification device. There are a number of different products on the market, available at most sporting goods stores wherever outdoor gear is sold, or through outdoor catalogs. Small hand-pump water filters are popular with backpackers, and for good reason. You will want something that not only filters the water but also purifies it. After any major disaster, available

water sources may be contaminated. Drinking water without first treating it is asking for trouble. It is to your advantage to invest in the best system you can buy, along with plenty of replacement filters.

The capability to boil water will also be important for killing harmful bacteria and preparing food. A mess kit pan or cooking pot, plus several methods for making fire, discussed below, should be included in your bag. A large metal cup is also an especially useful piece of equipment.

Making Fire
Systems for making and controlling fire are always high on the list of survival priorities. Manuals often recommend keeping at least three different types of fire-starter devices in the kit to ensure the ability to get a fire going even under adverse weather conditions. Methods may include flint and steel, matches, cigarette lighters, road flares, and magnesium fire strikers. Dry fuel, tinder, candles, and possibly canned fuels might be used in conjunction with an ignition system to sustain flames.

Fire can be extremely important for a variety of reasons, and I would encourage

Wood stoves could become lifesaving heat sources in cold climates.

every survivalist to master any of the more primitive methods for making fire, such as the bow drill or hand drill. Techniques for making fire will be discussed at length later in the book. Matches and lighter fuel will eventually be used up, but your need for heat will continue. Although emergency road flares might be considered somewhat bulky, I recommend storing several in your kit for the simple reason that they provide quick, convenient, reliable, and super-hot flames in any kind of weather. That's something that could possibly save your life in subfreezing or very wet conditions.

First-Aid

First-aid items, such as bandages, splints, thermometers, alcohol wipes, pain pills, antiseptics, and other miscellaneous emergency gear should find a place in your kit, as many of those things will be used eventually, if not right away.

Pre-assembled starter kits that contain a basic assortment of first-aid supplies can be found in many stores, including sporting goods stores and other discount retailers.

Important Tools

An assortment of hand tools might be inconvenient to haul around, but their utility value makes them well worth their weight. People will need tools to achieve certain tasks. You will want to pack some basic tools in your kit to help you repair things, build shelters, dig pits, skin and butcher animals, and accom-

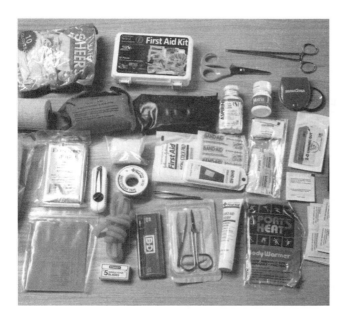

You could start with some basic first aid and medical gear and update the kit as time goes by.

A selection of basic hand tools for the long-term survival bag.

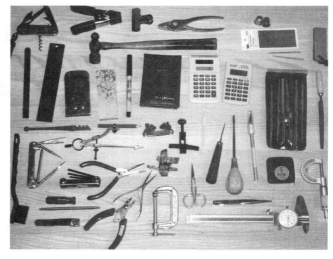

Small tools are sometimes indispensable, and they don't take up much space.

Access to conventional hardware stores could one day be severely limited, but miscellaneous hardware will always be needed.

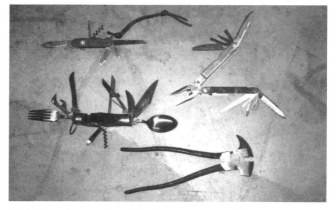

Multifunction tools might become especially valuable when people are forced to rely on their own resources to survive.

plish countless other chores. At the minimum, I would recommend a quality knife with sheath and sharpening stone, small to medium-sized ax, hacksaw with spare blades, small shovel or trowel, combination pliers such as the Leatherman Wave, sturdy scissors, any of the genuine Swiss Army Knives, leather awl, long pliers or tongs, a selection of metal files, and maybe a wood rasp, crosscut hand saw, and plenty of rope and cord. Space permitting, I would also include a curved hide-scraping blade, blocks and tackles, a selection of miscellaneous clamps, screw-drivers, adjustable wrenches, a pry bar, bolt cutters, tape measure, claw hammer, and a manual drill with an assortment of drill bits.

In addition to these main hand tools, you might add other small items such as manual can openers, razor blades, punches, hand wire, an assortment of screws, wing nuts, bolts, staples, nails, rivets, washers, roll and cotter pins, springs, safety pins, split rings, cable locks, paper clips, corks, handle wedges, duct tape, plastic lock ties, rubber bands, sewing needles and thimbles, and any other handy miscellaneous hardware you might think of. Paper and pencils will be useful, as will pocket calculators. A straight edge metal ruler might also be nice to have. A small bottle of machine or gun oil could become precious to you at some point.

You will have to give careful thought to how you pack everything. Blades and saw teeth should always be covered to protect other gear. Small things like screws should be contained separately from larger items to keep things organized and secure. You might have to experiment and repack everything several times to make things fit together for maximum space conservation inside the container. An oil bottle should perhaps be stored inside several sealed plastic bags, in case it spills. Any paper or wood products like pencils should also be protected inside plastic bags to keep them dry. Rubber bands can become quite sticky over a period of time, so they can go in their own little bag. Careful planning will make your kit a lot more usable when the time comes.

Such a bag of tools and hardware would constitute a significant expense for anyone attempting to purchase everything new. The majority of items mentioned here are typical of what I see sold at yard sales for pennies on the dollar. It is not uncommon to find a tool that normally retails in the hardware store for around $10–$12 being liquidated at the neighborhood yard sale for a quarter or 50 cents. Sometimes you have to clean away a layer of grease, sharpen a blade, or tighten up a loose hatchet head with a wedge, but it's a great way to outfit your kit with low investment.

Those who now regularly use tools in their occupations, chores at home, or hobbies will be able to appreciate how important they would be to survivors dependent upon their own capabilities for nearly everything. I would expect common hand tools to rank near the top of any survivalist's list of priorities for equipment. From buttering a slice of bread to pulling a cork out of a wine bottle, brushing teeth, shoveling snow off the driveway, scraping ice from a car windshield, trimming trees and shrubs, mowing lawns, changing car tires, and trimming fingernails, tools are a big part of human existence and always will be.

If you go to the trouble of setting up any supply caches at all, you might consider caching stores of extra tools and hardware for the purpose of bartering. You can bet your

money that tools and hardware will always be needed and coveted whenever industrial activities slow or cease. You might be able to trade tools for food, if you must.

Weapons

I am of the opinion that personal weapons such as firearms are essential in any harsh environment, for personal security as well as for hunting game. But again, we may visit this broad topic in the next Dark Age survival book.

I would include in the kit at least half a dozen quality bowstrings for the archery gear you may see fit to construct in the future when your supply of bullets begins to run out. I recommend purchasing strings of the highest quality and heaviest draw weight, and also the longest strings you can find, just to be on the safe side. If a string is too long for any bow you build, it might be made workable by twisting it a few times to shorten its length. A string can be made shorter to fit a shorter bow, but I don't know any way to make it longer. The longest and heaviest available strings might be the best survival strings. Of course, animal sinews have served as bowstrings for thousands of years, but you may not happen to have any on hand at the time when you need an expedient functional bow. Besides, the best modern bowstrings are superior to any made from natural material

Fishing Gear

Fishing tackle might be comprised of several hundred barbed hooks (#6 and #8 hooks are generally perfect for freshwater streams and lakes, and they don't cost much), several ounces of split-shot sinkers, a thousand yards of 10- to 20-pound test line, gill nets, at least a hundred swivels, and a mix of flies, jigs, spoons, plugs, plastic worms, and other lures. If you end up with extra tackle, you could always use some for barter.

Cordage

Cordage might be one of the most important products to the human race. The clothes

 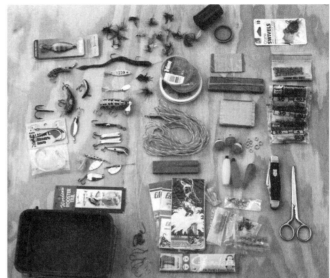

Fishing gear.

we wear are primarily woven from fine thread. Curtain drawstrings, shoelaces, clotheslines, dental floss, garden twine, water-ski ropes, fishing line, and flag and banner suspension cords are just some common examples of cordage being used in our modern world. In ancient times, cordage was used to suspend ship sails, sew cloth and hides for clothing, string bows for hunting and war, snare animals for food, construct fishing nets, rope bridges and ladders, and for binding things together, among possibly hundreds of other uses.

The modern survivalist has the advantage of being able to acquire manufactured ropes and cords, available in a wide variety of sizes and types of material, and in large quantity. There are plenty of books available to teach you how to tie knots and use cord in a multitude of crafts. Information will be provided in a later chapter on fabricating cordage from natural fibers and using a few basic knots.

Your need for cordage should be considered thoroughly when preparing for an uncertain future. I encourage readers to stock as much miscellaneous cord in their kits as possible to include strong threads, floss, fish line, parachute cord, and rope. Dental floss, both waxed and plain, might be

one of the most versatile types of small cord you could ever hope to have. In certain situations it has served as sewing thread, fish line, wrapping cord, snare line, and general-purpose string, in addition to its original purpose. It is very compact, inexpensive, lightweight, strong, and widely available as long as stores are open.

Another excellent product is #550 Paracord, or parachute cord. The genuine product consists of a nylon tube containing usually seven smaller individual strands, each a potentially versatile heavy thread. Genuine military-grade parachute cord is supposed to have a strength rating of up to 550 pounds, and it is about the same diameter as a bootlace. It comes in several colors, including olive green and black, and is usually available in military surplus stores and catalogs. At the very minimum, several hundred yards of this versatile product belong in your kit.

Considering the utility value of cordage, I might say that you could never really have too much of it. Fortunately, storing plenty of small-diameter cord is not a problem. You can wrap knife and ax handles or other gear with it, coil or stuff long pieces into small spaces, or lace it into the shoulder strap of your bag to help reinforce it. Parachute cord can be laced

Tool handles and gunstocks can be wrapped with miscellaneous survival cord to ensure supply.

into your hiking boots. There is really no excuse at all for not having a good supply of different cords when you need some. However, your supply could eventually become exhausted, so I would encourage readers to learn as much as they can about cord construction using different materials, as we will study later on.

Clothing

The clothing you decide to pack in your kit should be chosen for its durability and protective layers against the elements, among other things. Wool provides excellent insulation, even when wet, but it's heavier than certain materials. Also, not everyone finds wool comfortable to wear against the skin.

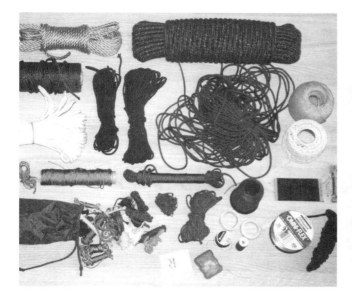

Now is a good time to start stocking up on miscellaneous utility cord, while it's available.

Functional antique treadle-operated sewing machines might become practical again someday.

Cotton is usually comfortable and light enough when dry, but it doesn't breathe well when wet or wick away perspiration the way wool does.

We have a lot of different choices for clothing fabric, with each material having its own advantages and disadvantages. The expensive outdoor wear commonly associated with mountain climbers and skiers might be desirable in extreme cold, if the usual bright colors can be tolerated. Whenever camouflage is desirable, modern military field clothing is the answer.

Clothing worn daily is subject to repetitive stresses and abrasions under normal use, and receives much more abuse in harsher conditions. Because you will want your limited supply of clothing to last as long as possible, ruggedness is a quality to specifically shop for. But even if you buy the best products available, your clothes will eventually wear out. In addition to packing some basic articles of clothing (several pairs of trousers, shirts, thermal underwear, maybe a dozen pairs of boot socks, boots and shoes, gloves/mittens, knitted cap or ski mask, cold weather parka, several sweaters, short pants for summer, and at least two sturdy leather belts per person) you should also include a comprehensive sewing/clothing repair kit. Needles and thread are cheap—buy extras. Repairs to clothing will eventually need to be made, and at some point you may find yourself constructing new clothing from old cloth or animal skins. Later in the book, these techniques will be discussed at length.

Personal Hygiene

Personal hygiene may become a real challenge in the next age. In parts of the world where they don't have enough water-treatment facilities or standard indoor plumbing, bathing is not as regular as what many of us are accustomed to. Anyone who has ever done very much remote camping can appreciate how the body feels after a week or two without a warm shower.

Small, sturdy plastic boxes might be ideal for storing assorted survival gear in the kit.

Hygiene and health are closely related. You need to keep your teeth and mouth clean to avoid tooth decay, abscesses, or gum disease. You need to keep your body clean to avoid myriad skin problems and other diseases. A complete personal hygiene kit could make a huge difference in your quality of life, at least until the soap and toothpaste run out. It is impossible to know how many toothbrushes you will need, but if you end up with extras, they might be valuable as barter items at some point. There will always be those who didn't think to purchase any extras when the stores were open.

Several washcloths might be useful. Toilet paper is always good to have. Fingernail clippers might one day be as valuable as quality scissors. A small mirror could be very important for this type of kit. Disposable shaving razors are better than nothing, but the old-style folding straight razors last a lot longer and are more versatile, though good ones are usually expensive and sometimes hard to find these days. Some type of skin cream, foot powder, and shaving cream might be helpful in the short term.

Money

Currency in small bills might be useful to people in certain short-term disaster situations. Gold and silver coins might be appropriate money after a collapse of our current economic system. This subject will be visited later in depth.

Miscellaneous

Other items that might be considered for the kit are blankets, sleeping bags, tents/tarps, thick plastic bags, binoculars or other optics (spare eyeglasses, if you wear glasses), lanterns or flashlights (plus a store of extra batteries), vegetable seeds, maps and compasses, insect repellent, chemical protective gear (gas masks), snowshoes, inflatable rubber rafts, usable reference literature, and anything else that might be especially useful to you or that you don't want to spend the rest of your life on Earth without. Any items of clothing, books, matches, or bottles of liquid should be sealed in waterproof bags. Optics and eyeglasses are best stored in either padded or hard-shell cases for protection.

Arrange your gear as you see fit, adding or subtracting any of these items to suit your own needs.

ENDNOTE

1. Amanda Ripley. "Floods, Tornadoes, Hurricanes, Wildfires, Earthquakes . . . Why We Don't Prepare." TIME magazine, posted Sunday, August 20, 2006. http://www.time.com/time/magazine/article/0,9171,1229102-1,00.html (Accessed 8-21-06)

CHAPTER 2

Caching Supplies

Imagine that you were forced to leave your house quickly and empty-handed, but you had this assortment of survival gear cached somewhere nearby.

People have been caching things throughout history for a variety of reasons. Treasure has been buried to keep wealth secret, contraband has been hidden from authorities, and supplies of all sorts have been concealed to retrieve at a later time, such as during a particular phase of an expedition or military operation. We will focus on caching for a future Dark Age—where survival gear of all kinds could be especially precious.

I recently retrieved a plastic cooler that I had filled with miscellaneous survival gear and buried in the woods five miles outside of town almost exactly seven years earlier. I'm not sure I would have been able to find it again without the written directions I made at the time, as my memory seemed to contradict my recorded notes. The importance of keeping an accurate record of the location was certainly reinforced in my mind. It turned out to be a valuable exercise. The contents of the cooler were found to be in perfect condition, confirming a good seal of the container despite my lack of confidence there.

Caching some of your supplies in different locations is one way to preserve your future ability to access provisions. You protect a percentage of your physical assets by employing a system of distribution, using multiple caches. If your home or storage unit were to burn to the ground or be burgled while you were away, you wouldn't lose everything. And, if you had to evacuate in a hurry to escape a natural disaster or other danger, whatever gear you've got in caches is something that you don't have to worry about transporting or dealing with immediately. If the caches are properly sealed and prepared and your reference map still applies after the disaster, your gear is probably safe.

Some survivalists use the strategy of creating two separate backups to their primary preparations. This might seem excessive in some cases, but making it a habit has obvious benefits; it is clearly the most effective preparedness strategy. Survival caching follows this methodology. Essential provisions and equipment, perhaps much of it obtained in duplicate, is hidden in different locations to ensure that a compromise of any single cache site will not compromise the entire supply. If you are part of a large group, you could employ a system whereby only a few members know the whereabouts of each individual cache location, thereby ensuring a certain degree of secrecy to perhaps a whole network of caches. A group could also spread the costs of the cached supplies among its members, minimizing the investment for each individual.

It should probably be noted at this point that the term "caching" is generally defined as the practice of storing or hiding supplies. This does not necessarily mean burying the supplies in the ground, although that might be the most common method used. A very temporary cache might be nothing more elaborate than simply hiding a pack of supplies under a mound of leaves or in a snowdrift. Caves have been popular hiding places for possessions for eons. Other options include basements, cellars, crawl spaces, wells, culverts, tunnels, bridges, walls, garden beds, and shrubs.

Things you might want to consider caching are coins and other currency, jewelry, tools, weapons and their accessories (such as ammunition, reloading components, spare parts and other equipment, brushes, solvents, oils, and lubricants), maps, instructional or reference literature, optics, hardware and raw materials, radios, animal traps, packs, clothing, tents, blankets, sleeping bags, food, fuels, candles, matches, batteries, medical supplies, vitamins, seeds, and even bottled water. Virtually anything you might need (or others might need and be willing to trade for) at some later time could be cached now to increase your survivability.

CACHING BASICS

After researching this subject and conducting some experiments of my own, I've discovered that it's easiest to keep the process organized and thorough by breaking it down into five major issues. If you give adequate consideration to each of these issues, your caching operations should be very successful.

1) Supply Selection: Deciding what will be cached
2) Protection: Providing for long-term protection and preservation of the cached supplies
3) Site Selection: Finding a safe and suitable location
4) Secrecy: Keeping the cache location and activity under wraps
5) Execution: Choosing the tools and methods for accomplishing the task

Supply Selection
Probably some of your choices of gear and supplies for your caches will mirror those that you've packed in your bags, but not all. In any case, you will be trying to ensure that your basic needs are met first, which alone could stretch the weight and volume limitations of your bags. Theoretically, caching could provide unlimited opportunities for supply storage. That's one of the more appealing

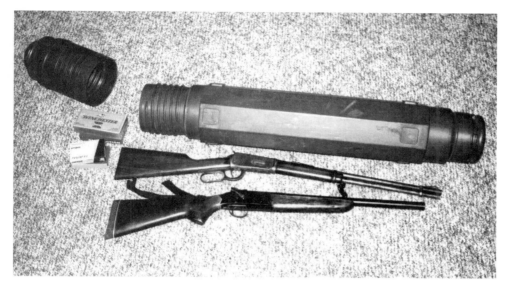

These two long guns spent a winter underground inside this plastic sonar buoy tube. They were recovered in the same condition they were in when buried.

aspects of caching. What you ultimately decide to hide in caches might depend in part on the size of available cache containers and on the tools and methods employed in the caching operation.

You may want to bury certain goods that you don't necessarily want to haul around with you. Canned goods, for example, often have a long shelf life and store well at a constant temperature, but are generally heavy and bulky. You may decide to store some cans of beef stew under the ground. Light bulbs may be worthless immediately following an event that causes a widespread blackout, but eventually some power systems might be restored. Radios were not mentioned in chapter one, as they would add bulk and weight to the kit that is probably already overloaded, and most require a supply of batteries. However, communications could eventually be very important to survivors, so radio gear might be a logical component of your cache supplies. Some would probably argue that certain radio systems belong in your main kit, and I wouldn't say they're wrong. You will have to establish your own program of prioritization.

Another thing to consider when selecting items for caches is how well they will stand up to the seasonal temperature variations in the ground at the depth you will bury them. In much of the world, the Earth's surface

temperature changes significantly from summer to winter. Naturally, the deeper the hole, the more constant the temperature. A shallow cache hole will be subject to a lot of temperature changes over a period of several years, which would affect some supplies more than others. If your packaging doesn't provide proper insulation and protection from the elements, you could easily lose a percentage of your supplies in the ground.

Protection

Once you've decided on exactly what you want to cache, you need to think about protecting it from the elements. The first step might be to select the appropriate container. The ideal container will easily house everything it is intended to house, as well as protect its contents from ground pressures and stresses, external moisture, and temperature fluctuations. It will be waterproof and airtight when closed, and it will be rot resistant, corrosion resistant, and sturdy. It will have the capability of being opened without damaging its contents.

Examples of containers may include capped sections of PVC pipe, sonar buoy tubes, mortar tubes, steel military ammo boxes, plastic buckets with lids, steel drums, glass jars, wide-mouthed plastic bottles (for small items), plastic pickle barrels, insulated picnic coolers, and other products specifically made for caching provisions. Some can be expected to work better than others: a glass jar might not survive rough handling, a slip with the shovel's blade, or drastic temperature changes in the ground without cracking. It's also comparatively heavy.

Before caching, supplies should be carefully selected, cleaned, dried, organized, packed, and, in some cases, individually wrapped or sealed to protect them from other items being stored in the same container.

How you cache your gear will depend a lot on how long it needs to stay hidden, the terrain, exactly what the supplies are and what they will be needed for, how secure you want the cache to be, and how much time you've got to do the job. Sometimes a shallow hole is advantageous because it requires less time and effort spent at the cache site—definitely something to think about. It might make sense to have different types of caches for different types of supplies. For instance, a weapons cache would be buried in a quality container and in a deep hole, while a convenient stash containing bars of soap for barter would receive less secure treatment.

Moisture

Moisture is one of the biggest enemies of long-term storage, especially in colder climates. This threat can be reduced with the use of desiccants.

A desiccant is a chemical product that essentially neutralizes the moisture problem in a confined space by absorbing it. The little silica gel packets often included with the packaging of new suitcases, cameras, and boots are a typical example.

Various types of desiccants are sold by such companies as Dillon Precision Products, Brigade Quartermasters, Hydrosorbent Dehumidifiers Company, Midway USA, Associated Bag Company, and Dry Pak Industries. Midway USA sells silica gel packs as well as rust-inhibitor chips.

To ensure that your container is absolutely waterproof, you could run a bead of

An insulated container like this plastic thermos would have obvious advantages in cold climates.

Typical large cache containers. The steel missile tube is heavy but will hold a lot of supplies.

An assortment of small metal and plastic containers that could be used for caches.

ABS pipe, such as this 4-inch diameter pipe, is not expensive. Everything needed for this medium-sized container was purchased at a hardware store for around $20.

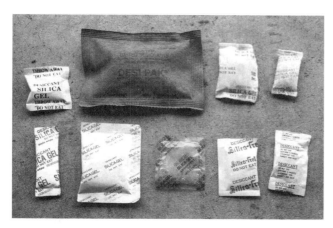

Desiccants are used to absorb moisture within a confined space.

silicone bathtub caulk over every seam or junction. If your container is PVC pipe, a special glue is available.

Because most desiccants absorb and retain moisture, you won't want them in direct contact with anything you are trying to protect from moisture. You could store firearms and other items in individual gun socks to provide a barrier between them and the desiccants, or you could keep the desiccants contained within their own sock or breathable bag inside the cache container. Desiccants also have a limited lifespan. I wouldn't count on them making a huge difference much beyond a period of several months.

Food

Vacuum-packing your supplies could be a useful method for underground storage. Cabela's routinely offers the FoodSaver Vacuum Packing System, which will vacuum-seal foods and other small items for long-term storage. It might also prove useful to browse the Internet for other heat-sealer products available through packaging, canning, and food processing suppliers. Depending on what you find, vacuum sealers generally cost anywhere from a little over $100 to more than $300.

Food products should never be exposed to desiccants. Food-safe oxygen absorbers, available from Dry Pak Industries, prevent mold and mildew from forming and aid in long-term storage.

Firearms

For reasons of safety, weapons should be prepared and stored unloaded. Ammunition stored in the same container with the weapons could be separately wrapped or sealed.

When preparing firearms for caching, special precautions should be taken to preserve your anonymity should someone else discover the cache. Prior to caching, firearms should be handled with rubber gloves and any fingerprints should be removed, both to prevent the guns from being traced back to you as well as to prevent surface rust on the steel. You should only cache guns that cannot

This 20mm ammo can is made of heavy-gauge steel with a rubber gasket in the lid that forms an airtight seal. It is a good cache container for ammunition and handguns.

be linked to you—guns with no paper trail, such as those you acquired without signing forms, perhaps purchased from non-dealers in another city. You never know who might someday find your cache; it is best not to have any links to it.

The tough plastic sonar buoy tubes seen regularly in surplus catalogs a decade ago make excellent cache containers for long guns and other gear, but seem to be rather scarce lately. A decent tube container can be made from ABS or PVC plastic sewer pipe of adequate diameter. Check with your local plumbing supplies dealer. For most long guns, you will want a tube with a minimum 6-inch inside diameter. End caps can be glued on; but make sure one end has a female adapter with a threaded screw plug for easy opening. PVC products are very good, but expensive. If you can afford it, a 3- or 4-foot length of 8-inch diameter PVC tube at around $4 per foot, plus end cap ($20), threaded female adapter ($57), screw plug ($42), and the PVC cement, altogether will make a fine cache container for long guns and related gear. This size of pipe is sold in 13-foot lengths in my area, so it might make sense to buy everything you need to build several containers at one time.

If firearms must be disassembled to fit into the container, make sure that all of the parts are kept together. Assembly/disassembly instructions may also be helpful, kept with the weapon in a zip-locked freezer bag. You will want to remove all gunpowder residue and any excess oils or solvents that would trap moisture.

William Nelson and Stanley Catlow discuss Cosmoline, a rust-preventative grease used to protect metal parts, in their interesting pamphlet, *Methods of Long Term Underground Storage* (available from Delta Press and other preparedness sites). Packing guns and other metal objects in grease for storage has been a common practice for many years, but it wouldn't be my method of choice for several reasons. I don't like the idea of having to remove the Cosmoline when I recover my gear—it's messy and time-consuming, and could be extremely inconvenient when time might be critical. Also, I've seen plenty of greased metal objects that have rusted anyway, since grease can actually trap moisture. However, I should confess that I have no personal experience experimenting with the grease method of preservation. If you think it might work for you, Brownells sells Cosmoline and other rust preventative products.

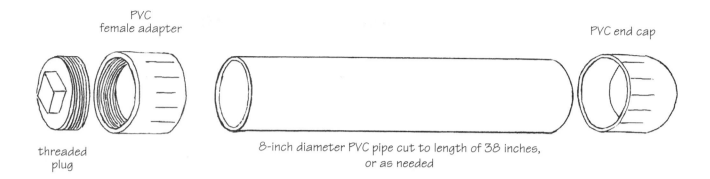

PVC
female adapter

PVC end cap

threaded
plug

8-inch diameter PVC pipe cut to length of 38 inches,
or as needed

The perfect cache tube. Female adapter and end cap are glued on to the tube body using special PVC cement.

Before leaving the subject of protecting your underground cache, a more elaborate method might be worth considering. Filling the container with an inert gas like nitrogen can retard the decaying and corroding processes and would be especially effective in preventing rust on firearms and other metal products. You would need to install a special valve opening on the container and fill the inside with the gas obtained from an industrial supplier to a prescribed pressure. This requires additional time, effort, and cost. For more on this and other sophisticated caching techniques, I recommend the excellent video, *State-of-the-Art Survival Caching*, available from Paladin Press.

Site Selection

Assuming that you've selected a general area for your cache site, your first task might be to scout the area and learn as much as you can about it. You will want to know about any roads or alternative routes in and out, the normal level of human activity, the hardness of the ground, and whatever permanent fixtures you might find to serve as landmarks and reference points. You can probe the soil with a metal rod to test its consistency, and then choose digging tools accordingly. You could make a preliminary study of the environment prior to caching any supplies. Consult maps and take plenty of notes.

When selecting your site, be mindful that low ground could flood, mountaintops may expose you to high visibility, and any commercial or private property may be subject to future development. Consider possible mud slides on open hillsides. Caching your valuable supplies should encourage you to be forward thinking.

A topographic map of the area could be very useful, along with a decent orienteering compass and a protractor. Locating your position on a map from compass bearings (azimuths) to two or more visible terrain features shown on the map is called *resection*. Any measurements you make for reference will warrant clear and accurate written notations. Human memory is generally unreliable with details over a long period of time; it is advisable to double-check all readings to verify accuracy. The only possible drawback to having a written record of your cache location is that it could find its way into the wrong hands and compromise your efforts.

Another possible method of recording locations of cache sites is with the use of GPS (global positioning system). This technology might be useful, but keep in mind that a GPS receiver is battery-dependent and relies on satellites in space. Its long-term usefulness after a collapse of modern civilization could be questionable.

A typical secluded area possesses plenty of possibilities for caching supplies.

Binoculars come in handy when scouting terrain for cache sites and for surveying an area to monitor any human activity.

In any case, I prefer a combination of reference systems. You will want multiple landmarks for reference. Trees are OK, except that a forest fire or a progressive logging operation could alter the scenery substantially. Distant mountain peaks or large rocks are excellent for shooting compass bearings to, when they are visible, because they are permanent. You can also create your own reference markers, but they should be subtle enough that others won't notice them, or at least be able to interpret them correctly.

Secrecy

After you have chosen your site, you must address secrecy. You are hiding certain gear from others to keep it secure until needed. Exactly how, when, and where you cache your supplies will be determined by a number of variables, such as the type of supplies, the size of the cache, the terrain, the season and weather conditions, the available time within which to accomplish the task, your access to remote or secret locations, the number of people involved, and any number of other things that might influence your decisions and methods. Some might choose to travel far or backpack deep into the backwoods, while others might find ideal places within the boundaries of their own property.

Probably the best argument against hiding or burying your goods on your own land is that your legal property, especially your private residence, presents a more confined search area for anyone else hunting for your cache. If someone happens to be looking for what you have to hide, your own property is a logical first place to search. It might also be difficult to avoid nosy neighbors unless you're fortunate enough to be on rural property outside normal public view, or have property surrounded by a thick grove of trees, high rock wall, or solid fence. Also, if you were forced to evacuate your premises in a hurry, you could lose immediate access to any supplies buried there. All things considered, your own property may still offer some good possibilities, and you may not find a more convenient location. Again, use your imagination.

The best way to keep your cache sites and activity secret is to operate alone. The less information others have about what you are doing, the higher your level of secrecy. Even so, involving another person has certain advantages. Several people can share the workload. A well-rehearsed team can sometimes work more efficiently than someone working alone. One person could act as a guard or lookout while the other concentrated on the digging. Several people can share ideas, and the whole task could be speeded up by

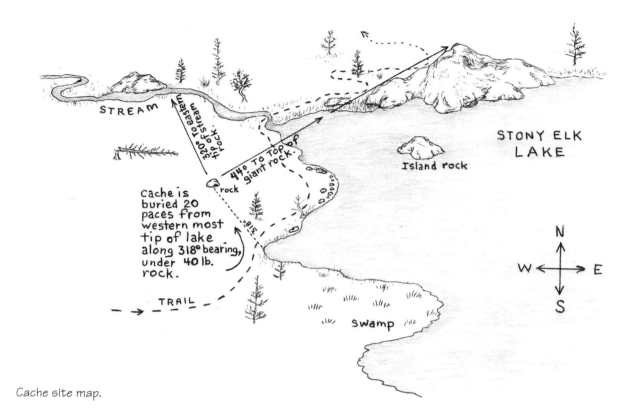

STREAM

320° To eastern tip of stream rock

44° To Top of giant rock

Cache is buried 20 paces from western most tip of lake along 318° bearing, under 40 lb. rock.

rock

318°

Island rock

STONY ELK LAKE

TRAIL

Swamp

N

W ← → E

S

Cache site map.

sharing the work. Consider that the shorter the amount of time spent at the cache site, the less chance of being discovered. For this reason, always fill and seal your containers *before* going to the cache site. You want your package ready to drop into the hole to save yourself time and hassle at the chosen location. You will have plenty of other things to be concerned with there.

Ideally, the spot you've chosen is a relatively secluded area hidden from all view, preferably outside the normal audible detection range of other people. The shovel will clank and ring if it strikes rocks, but you should endeavor to be as silent in your efforts as possible.

If aerial surveillance by satellites or aircraft is a concern, look for overhead cover at the cache location. This might be provided by a thick stand of trees, a bridge, a cave, or other natural elements. Think about your side visibility as well. It is imperative that your caching activity is concluded without being observed by anyone. (If you saw someone

burying something in the woods, wouldn't you be tempted to investigate the scene after the person left?)

While excavating your hole in the ground, it might be wise to periodically halt all activity for brief spells to quietly listen to the environment and maybe scan your surroundings with your binoculars, just to confirm your privacy. Maybe nobody else was in the area when you started digging, but as your work progressed some hikers wandered up the trail and heard you chopping up the ground. You don't want to become so preoccupied with your task that you fail to notice others nearby.

The proper clothing to wear for this activity would be comfortable outdoor work clothes—no bright colors. They should also be appropriate for the season.

If you research this subject enough, you will encounter plenty of ideas about defeating metal detectors. These include scattering junk metal about the cache site in order to confuse the search coil, burying caches under obvious metal objects and structures, and even tailoring a specific cache so that it doesn't have any metal content. We won't focus on this potential threat here, since a totally secret cache site in a remote setting is not a likely place to find someone with a detector. Before any searcher would use a detector, he would first have to have some idea about where to conduct his search. The mountains, parks, national forests, deserts, and other public lands in America are perhaps all too vast to ever be thoroughly searched by metal detectors, at least in your lifetime.

Execution

Finally, you need to think about carrying out the task of hauling your cache to your hiding spot, burying it in an appropriate hole or trench, and leaving the area without being seen or leaving telltale signs of your activities.

Choose your caching season carefully. In many states during winter, the ground is often either frozen hard or covered with snow. Hard ground obviously makes digging more difficult, and it might be impossible to avoid leaving tracks in snow. During hunting season, usually in the fall, you would run the risk of encountering or being observed by the numerous hunters in the woods. If by chance your activities are discovered, what will your story be?

A camping trip might provide an excellent opportunity for caching your supplies. Things like shovels and large containers are expected around campsites, and certain activities, such as digging holes in the ground, might seem less suspicious around a camp.

A large weapons cache might be too heavy for one person to easily carry very far. Transporting it to the chosen cache site might require the aid of a 4x4 vehicle, a makeshift travois, or the help of trusted friends or relatives. Figure all of these issues out carefully ahead of time, then execute your task without taking shortcuts, and you'll likely sleep better at night.

It might be helpful to have a backpack suitable for carrying needed tools to the site, as well as the container itself, if it's small enough

to fit into or be strapped to a backpack. I've used an old, canvas, foreign army pack, and it has served me well for this purpose.

Recording Your Location

Measuring tools are almost essential for accurately recording the location of your cache. At the minimum, you'll want a tape measure. I like a 100-foot tape. A length of cord marked with tape wrappings at intervals might be useful for swinging a radius or measuring a line distance. Include paper and pen or pencil for making notations.

Maps of the area, especially large-scale topographic maps, are handy. You may want to sketch your own maps as well, in conjunction with your written instructions detailing how to find the cache site. A decent map is made even more useful with a good compass. One of the best civilian-type orienteering compasses available is the Silva Ranger. The genuine U.S. Army lensatic compass with self-luminous Tritium is available from Brigade Quartermasters. Imported copies are a lot cheaper, but also inferior. Any quality compass should serve your needs here, as long as the Earth's magnetic pole doesn't change drastically.

A laser range finder, if your kit (and your budget) has room for one, could be very useful for determining distances accurately and conveniently from a stationary position. The thing to keep in mind when relying on these gadgets with sophisticated technology is that you may not be able to count on having them all available to you later when you need to retrieve your supplies. This is perhaps the strongest argument against relying on any single referencing system.

Digging

You'll need a small shovel, garden trowel, or military trifold entrenching tool. The trifold shovel is handy because it can be conveniently stowed folded in its own protective cover within your pack. The American-made products are durable and generally cost around $30–$45. Imported copies are cheaper, but definitely inferior. I've

Caching checklist: backpack, small tarp to catch dirt, map of area, compass, pocket calculator, notebook and pen for recording details, probe rod to test ground hardness, 100-foot tape measure, trowel and shovel, water bottle or full canteen, gloves, binoculars, towel, toilet paper, container with cache supplies sealed inside, ready for burial.

found a short, fixed-handle camp shovel useful and easy enough to carry with the rucksack.

Survival author Ragnar Benson has written about using a post auger for boring deeper holes. Keep in mind, however, that a full-size shovel or auger will be more difficult to conceal if you travel on foot to your cache site, and you might draw unwanted attention to your activities if you are seen carrying a shovel into or out of a wooded area. Although burying your cache tube in a vertical position may make it more difficult for others to find, it would also require a deeper hole than if you were to bury it horizontally. It could also make it more difficult to retrieve, especially if it is a large tube packed with heavy firearms and ammunition.

You might bring along a probe rod for testing ground hardness to have an idea about what you're getting into *before* you start digging. Probing the ground will let you know if there are any sizable rocks or tree roots where you want to dig, which would slow you way down. I use a round spike about a foot long with a 3/8-inch diameter. (I keep the pointed end capped with an empty .38 Special casing while it's in the pack.)

Other Tools and Accessories

A small tarp or heavy sheet of plastic is useful for heaping the dirt onto. This will make it easier for you to restore the area after you're done, because you'll scatter less dirt around. It also speeds up the job of filling in the hole.

Binoculars are important for visually scanning your surroundings. If any other people are in the area, you'll want to know about them before they know about you. Try different models before you purchase. The ideal model is easy and comfortable for you to use and is rubber-armored or shock-resistant, waterproof, and compact, and has quality optics.

A small flashlight will come in handy if you conduct your business at night, early morning, or late afternoon. A good night-vision optic might also be useful if you plan to operate in the dark.

Since you'll be getting dirty, you'll want rags or paper towels for wiping most of the dirt off yourself. Individually wrapped alcohol wipes or baby wipes might be useful for quick cleaning of your hands and face. A pair of leather work gloves might save you from blisters and dirt under the fingernails.

One of the most important things you can bring with you is a full canteen or bottle of water. If you don't get thirsty while hiking to the location, you surely will after you start digging.

Take an inventory of your gear used at the site before and after you leave to avoid leaving anything behind. Restore the site to the way you found it—leave no trace of your activity.

CHAPTER 3

The Survival Workshop

The ability to manufacture tools may become necessary for your survival.

One of the first priorities of survivors in a future Dark Age, after the short-term medical and health concerns are dealt with, of course, might be the production of the tools and hardware needed for long-term survival, rebuilding infrastructure, and a new system of commerce and trade. New products of all kinds would be needed, and the manufacturing of products requires tools.

The ability to transform metal scrap or raw material into usable products could become essential, and the survivor with the skills to make tools might find himself quite in demand. Having tools to barter for food and other necessities could make all the difference.

So many common things that we often take for granted—such as cars and appliances and even household hardware such as door hinges and deck screws—contain metal. Sheets of paper are held together with metal paperclips or staples; we cook with metal pots and pans in our kitchens; and even the wiring and plumbing inside our homes contain enormous amounts of metal. Imagine trying to get by in a world without any metal at all. While it is true that a growing number of traditionally metallic products are now sometimes made of plastic or other newly developed materials, modern societies still depend heavily on metals for numerous purposes.

I would guess that a working knowledge of metallurgy, the science and technology of metals, would be extremely valuable to survivors after a worldwide collapse of civilization. (This is true now, given the value of steel, aluminum, copper, and other metals.) Even if it were possible for a society to revert to a Stone-Age technological level almost entirely, enough scrap metal exists on the planet to provide some people with metallic tools, and there would most likely always be a percentage of survivors with some knowledge of mining and processing metal. Mankind would have to be totally eradicated in order to completely erase metal technology, I believe. For the tools in a survival workshop, iron and steel will probably be our most important materials.

I've found that most kinds of common hand tools are not particularly difficult to fabricate. Perhaps the most challenging tools to create in a home workshop would be hacksaw blades and fine-cut files; survivalists might be wise to stockpile miscellaneous files and saw blades along with plenty of raw materials. Most everything else, from hammers, punches, and chisels, to pliers, shovels, knives, and axes, can be hammer-forged in a blacksmith's shop or ground out from other broken tools.

Ideally, the survivor's metal shop will be set up to handle both the cold and hot working operations. If electricity were available, sophisticated lathes, grinders, power hacksaws, milling machines, and drill presses could be wonderfully useful in the shop. In some situations when the neighborhood power is out, it might be practical to run power tools with a gasoline-powered generator, if the precious fuel can be spared. If power tools are out of the question, quite a lot can still be accomplished with manually operated tools.

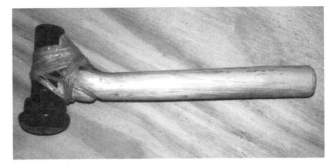

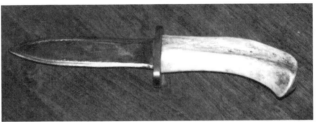

Makeshift handles restore purpose to these metal tools.

A hand-crank grinding wheel for those axes to grind when the power is out.

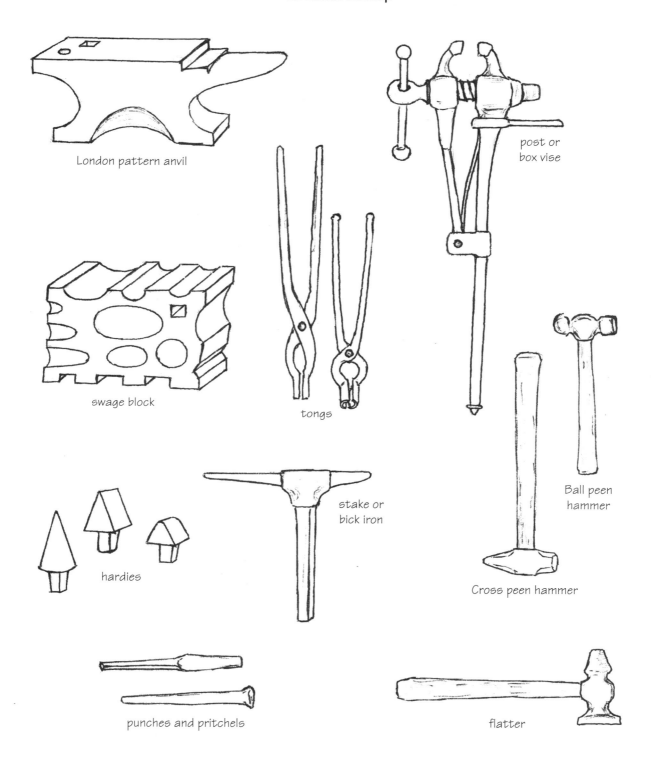

London pattern anvil

post or box vise

swage block

tongs

hardies

stake or bick iron

Ball peen hammer

Cross peen hammer

punches and pritchels

flatter

Basic tools of the blacksmith.

IRON AND STEEL

Iron is an element (atomic number 26, symbol Fe), and therefore is a pure substance composed of only one kind of material. Steel, on the other hand, is comprised of several different elements, including iron and carbon and sometimes other additional elements or alloys.

Genuine wrought iron is a material that is rarely used anymore but was once very common. It contains almost no carbon and is very soft. It has excellent corrosion resistance and is easier to forge weld than most steels because of its lack of carbon and its high silicon slag content. Low-carbon and low-alloy steels have replaced wrought iron as a common structural material in most applications.

Cast iron is iron containing more carbon than the amount contained in high-carbon steel, usually between 2.0 and 6.0 percent carbon. Unlike carbon steel, cast iron's carbon is no longer in solution; in other words, the amount of carbon is too high to be completely dissolved into the iron. Cast iron is very different from pure iron or steel, being very brittle and having low strength in tension, but with good strength in compression and having good wear resistance. Cast iron is normally easier to cast and machine than high-carbon steel. Several varieties of cast iron are popular for use in engine blocks, brake drums, cooking vessels, and other special applications.

Metals containing iron are called ferrous metals, characterized by their magnetic quality. In addition to cast iron and carbon steels, various alloy steels are commonly used, including tool steels, stainless steel, spring steel, and other special-purpose alloys.

There are numerous alloys of steel, each having its own special characteristics. Most steels contain more than 90 percent iron, with the percentage of carbon commonly ranging somewhere between 0.15 and 1.0 percent. As a rule, the higher the carbon content, the harder and more brittle the material can be. Low carbon, or "mild" steel, is more flexible and ductile than high-carbon steel and cannot be hardened by simple heat treatment.

The vast majority of the steel produced in the world today is soft, low-carbon steel, which is used in car bodies, street sign posts, chain-link fences, nails, paper clips, and literally thousands of other products. High-carbon steel and other hard steel alloys are typically used in cutting tools that will hold an edge, or in machine gears that require high wear resistance.

Steel ingots are marked to indicate their type or alloy content. Of course, it is highly unlikely that any piece of metal found in a future scrap pile would be marked this way, so in many cases it might be difficult to determine the exact alloys present in the materials encountered. It is nevertheless useful to understand the characteristics and practical applications of the various different types of steel and to have a referencing system.

High-carbon steel can be tempered, and is a suitable material for a variety of cutting tools. There are ways to test steel to determine whether or not it has high carbon content. When held to a grinding wheel, high-carbon steel creates tiny, bright yellow or white star-like sparks, whereas low-carbon steel creates longer, duller orange sparks. Also, after heating the material to a light cherry red glow, quenching it in water and then attempting to file it, the file will tend to slide over the surface of high-carbon steel, but will easily cut into unhardened low-carbon steel.

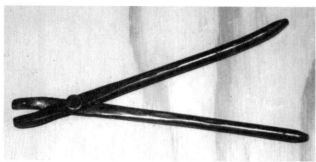

Pliers and tongs are useful tools that can be fabricated from many different materials. Top, pliers made from 3/8-inch round bar; middle, tongs made from rebar; and bottom, a small set of pliers made from two large nails, which uses a smaller nail as a rivet.

A multitude of tools and hardware can be created in a low-tech blacksmith shop.

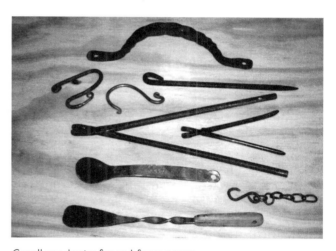

Small products forged from scrap.

Almost anything that can be envisioned can be created by someone with the imagination, patience, and determination to see a project through its stages. I do a lot of things by trial and error, sometimes needlessly perhaps, but I view the trial and error failures and successes as valuable elements to the learning process.

BACK TO BASICS

It seems logical to me that the best survival workshop would be one that does not depend primarily on electricity, because we might not be able to count on having reliable electricity for a period of time following a severe global crisis. Also, I believe the best survival workshop should be set up for metal working, in at least part of the shop. Tools used in metalwork, such as a forge and anvil, bench grinder, hacksaw, and lathe, can all be used to make other tools. What you're setting up is basically a blacksmith's shop.

A very simple shop organized for metal work should not be terribly difficult to set up under most conditions, even with very limited available resources. The basic necessary components include a forge or other heat source, anvil, hammers, tongs, and ideally a hacksaw and a selection of files. Some type of large vise, grinding wheel, water bucket for tool quenching and controlling fires, plus a selection of forming and cutting hardies, swage blocks, punches, and other blacksmith tools can be very useful as well. With just these basics, an endless array of tools, weapons, hardware, and other products can be manufactured. In fact, the tools used by the blacksmith can themselves be made by the blacksmith.

Thick leather welder's gloves and a leather apron are useful protective gear for any blacksmith, as is a pair of goggles or eyeglasses. Especially in the higher temperature ranges used in forge welding and foundry work, molten metal and hot slag tend to splatter. I would also recommend wearing thick leather boots when working in the shop. The floor of any blacksmith shop should be made of dirt, gravel, brick, concrete slab, or anything that won't catch on fire.

Arranging the Shop

In my experience, most of the forge tools are fully usable only when they are within arm's reach of the forge. If the smith has to step across his shop or around other tools to use the swage block, the anvil, other forming tools, or the vise, his work will likely already have cooled enough to significantly reduce its malleability by the time the hammering or bending operations commence. The best layout for the blacksmith shop in my opinion is with all of the necessary tools arranged so that they are easily accessible to the blacksmith as he stands in one station, close to the forge.

Dim light is normally preferred over bright light during hot metal work, in order for the smith to accurately judge the various heat stages by the glowing colors in the metal. A candle or kerosene lantern might be all the lighting such a shop would ever need, and even then probably only while working at night.

A bucket half full of water is useful for quenching hot metal as well as for putting out accidental fires if sparks ignite shop rags or clothing. I also keep a fire extinguisher mounted on the wall in my blacksmith shop. Wearing protective clothing like leather aprons, leather gloves, work boots, and protec-

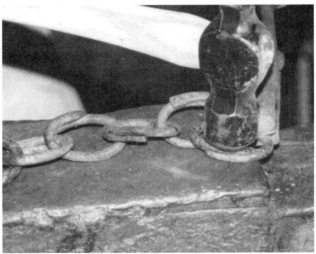

Above, making chain from mild steel.

Chain link successfully forge-welded closed.

Wood is not the ideal material for a forge frame, but it works here as a temporary expedient with this homemade contraption.

An indoor forge should have a chimney over it to minimize problems with smoke and soot.

Firing the coal in the makeshift forge.

tive eyeglasses as previously recommended could prevent a lot of painful burns. Chunks of metal that have been heated in the forge will hold quite a lot of heat long after they've lost their glow. Pieces that have fallen on the floor should be either avoided or handled with tongs.

The Forge

I think of the forge as the nucleus of the metal shop. It is where the metal is heated so that it can be worked. Without a heat source such as a forge or a gas torch, metal has to be worked cold, which limits you to sawing, filing, screw cutting, grinding, drilling, milling and lathe cutting, surface engraving, and some cold bending. These processes are often limited to mild steel or annealed harder steels, because tempered high-carbon and tool steels are very difficult to cut. Hot metal, on the other hand, can be drawn, twisted, bent, flattened, tapered, thinned, curled, upset, punched, cut, welded, tempered, annealed, case hardened, compressed, and formed into numerous shapes with hammer blows. Iron becomes malleable when sufficiently heated, changing to a consistency somewhat similar to wet clay. In the upper heat ranges, molten metals can be cast into complex parts. These processes can all be accomplished with the help of a coal forge.

Building a Forge

A Dark Age workshop might be low tech in many ways but still quite versatile. A forge of one style or another could give a craftsman or handyman enhanced capabilities, as noted above. Essentially, a forge is a bowl or fire pit usually containing either coal or charcoal for fuel to which forced air is supplied to raise the temperature of the fire. In ancient times, this function was achieved in earthen pits lined with stones or clay and fed with air pipes. Drafts of air were supplied to early forges with human lung power, with small fans waved by hand, and at some point in history with air-pumping leather or canvas bags called bellows.

A usable forge can be a very simple arrangement consisting basically of some type of bowl for the coal and some type of blower hooked up with a pipe to supply air to the fire in the bowl. Makeshift forges have been made in modern times using discarded hot water heaters, tire rims, washtubs, barbecue grills, steel pans, drums, wood stoves, and other items adapted with ingenuity. In a primitive situation with severely limited tools and resources, a pit in the ground might serve as the basin. A more elaborate stone or masonry fixture might be desired in a permanent shop. If some type of thin steel bowl is used, such as a common backyard barbecue, the inside should be lined with firebricks or clay. I built an experimental forge with a plywood box lined with concrete, and it works.

Whenever working around fire, especially in a confined area such as a shop with a coal forge, a chimney and hood are essential. I tried using a small open-bowl forge without a hood in my tiny blacksmith shop, since it is well ventilated with its large doors open, but quickly realized the necessity of a hood over the fire. Soot and smoke will fill a small shed in short order, making the working conditions unbearable. A simple sheet-metal hood shaped like an upside-down funnel over the bowl, connected to a section of stovepipe plumbed through the roof, solves most of the problem.

Fuel

Charcoal was used to fuel forges for thousands of years. It is still preferred by some metalsmiths even today. Charcoal is just wood that has been completely charred. It can be produced by cooking chunks of wood contained in a steel canister with a lid latched closed, with only a small opening for gases to escape. The heated wood, being mostly starved of oxygen, retains its structure and turns to carbon rather than burning up into ashes. Smoke and gas will emit from the vent as the canister heats up in the fire. The canister should be taken out of the fire when flames begin shooting out of the small opening. At that point the wood is usually sufficiently charred.

The best wood for making charcoal is a matter of opinion. Some prefer hardwoods while others prefer softwoods. Charcoal briquettes for backyard barbecues are generally not considered suitable for firing a forge, mainly because they would burn up too quickly. However, if briquettes were the only fuel source available, certain basic forge operations might still be accomplished.

Perhaps more popular with blacksmiths in recent times is bituminous soft coal, or forge coal with anthracite, because a high heat can be reached with it fairly easily. I buy my coal at a farm store near my home. Coal is the dirtiest-burning fuel I know of, but it works well.

Gas-fired forges are preferred by some smiths who desire a cleaner fire. As long as a

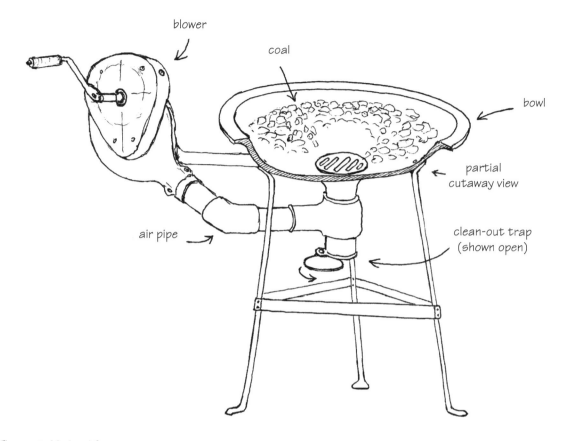

blower

coal

bowl

partial
cutaway view

air pipe

clean-out trap
(shown open)

The portable bowl forge.

supply of gas is available, a gas forge might be desirable wherever sparks, smoke, or the filth of a coal or charcoal fire are a problem. In my view, the biggest advantage to using charcoal, rather than gas or coal, in the forge is that charcoal is a fuel that can be produced by the blacksmith wherever wood grows.

Air Supply

Forced air can be supplied to the fuel by different methods. Bellows, sometimes called lung bags, were used for hundreds of years. The hand-cranked forge blowers popular in the early 1900s provided a steady flow of air and were very convenient to use. Old forges with crank blowers are occasionally found at farm auctions, flea markets, antique stores, and at rural yard sales. Electric blowers are still manufactured for different purposes and can normally be used to supply air to a forge.

Electric hair dryers and vacuum cleaners have even been adapted to supply air to forges. I constructed a small single-lung bellows system for an experimental forge, and while it does pump air successfully into the bowl, it requires a lot of work pumping the handle to keep the air supply constant. I imagine that a larger, double-lung system would work a lot more efficiently. My favorite system with the coal forge is the Champion forge blower, cranked by hand.

If a ready-made blower is not available, some type of functional bellows might be devised. I've seen bellows with the bag shaped like a tube that is pushed and pulled like an accordion. The more common teardrop shape has one or two hinged leaf sections made of wood and skirted with canvas or soft leather that fan apart and then close together to force air through a nozzle. A very simple small version that has only one lung and one

Old Champion blowers operate with a hand crank and can supply steady air to the fire in a forge.

Examples of various small air-supply devices.

hinged flap is a popular item in some homes among the fireplace tools. The traditional forge bellows were larger and generally contained more than one lung chamber, making them more efficient.

The basic principle of a bellows entails a system of lung bags connected to some type of rigid framework such as wooden flaps as described previously, with air chambers and air valves such that compressing the lungs forces air through a pipe or nozzle. The valves function by allowing air into the lungs when opening the flaps (expanding the bag), and preventing the air from escaping through any opening besides the nozzle as the flaps are closed (compressing the bag).

The simplest type of air-intake valve for bellows consists of a soft flap, such as a piece of leather, positioned over an air hole on the inside of the chamber in such a way that the pressure against the flap occurring when the bag is compressed forces it flat over the hole to prevent air from escaping. The suction created by expanding the bag pulls the flap open, allowing air in.

In my experiments, I have found that an air-intake hole that is too large interferes with the compression and suction forces that make the valve work properly. I eventually determined that one air hole the diameter of a pencil was ideal for the size of bellows I made, rather than the two 5/8-inch diameter holes I initially drilled through the hinged leaf. Also, I learned the hard way that the leather valve flap needs to be quite soft in order to yield properly to the forces that make it function. My bellows weren't pumping much air at first, and I eventually realized that the fairly stiff flaps I used weren't doing anything when the bellows were being pumped. Once I corrected these simple matters, the whole system worked as intended.

I also built a nozzle chamber with a one-way air hole to prevent the bellows from sucking hot ashes and soot into the bag when expanded. Air should only enter the lung at the intended air-intake hole, positioned several feet away from the hot coals.

I've seen bellows bags made of canvas and also made of leather. I used canvas painter's drop cloth for my bellows, but I would expect some type of cloth with a tighter weave to function more efficiently. Leather is probably the best material for this purpose, and it could

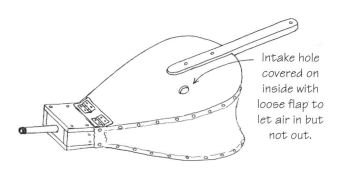

Intake hole covered on inside with loose flap to let air in but not out.

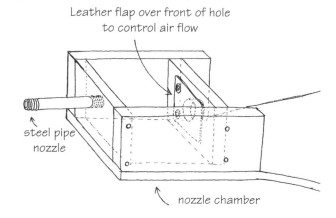

Leather flap over front of hole to control air flow

steel pipe nozzle

nozzle chamber

Homemade single-lung bellows to supply air to a small forge.

The nozzle chamber of the single-lung bellows, showing the one-way air valve.

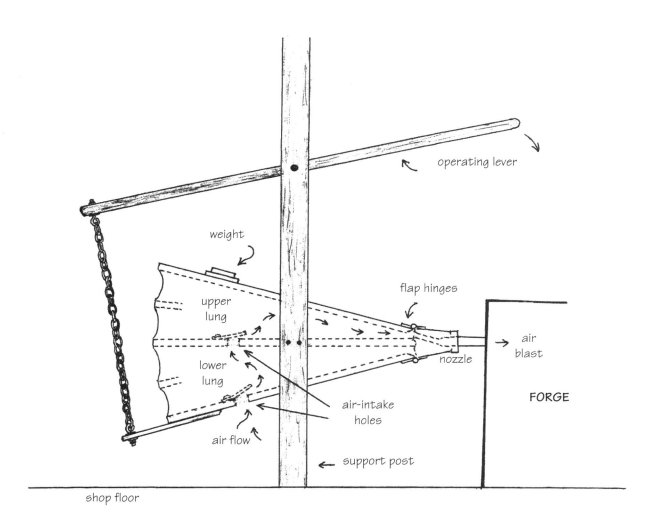

operating lever

weight

upper lung

flap hinges

lower lung

air-intake holes

nozzle

air blast

FORGE

air flow

support post

shop floor

Double-lung bellows.

be maintained with neat's-foot oil to prevent drying and cracking. I used a strip of leather under the hinges to prevent air from leaking out of the seam between the boards. All stationary seams and cracks where air might escape were caulked with hide glue, as that was what I had on hand during the project.

A double-lung bellows has two diaphragm chambers and provides a more regular flow of air if built properly. In the best arrangement, compression of the lower chamber (lifting the bottom flap) forces air into the upper chamber and, ultimately, through the nozzle of the upper chamber. The top paddle or flap is usually weighted, and downward pressure on this continues compressing the upper chamber as the lower leaf drops down, continuing the process of forcing air through the nozzle even as the lower chamber expands to take in air. A good double-diaphragm bellows alleviates any need for a separate nozzle chamber that would prevent air from being sucked back into the bellows from the hot forge, because the top lung, from which air escapes only through the nozzle, is pushing air when the operating lever swings both up and down with this double-action process. This system keeps the air flowing from the nozzle more regularly than what is possible with the single-diaphragm system.

In the book, *The Complete Modern Blacksmith*, author Alexander Weygers illustrates an interesting system for pumping air into a forge in a blacksmith shop he visited in Java. His illustration shows a forge with a clay hearth and two bamboo pipes protruding upward from the clay hearth at one end, and a small boy sitting atop a stand where he is able to pump reed pistons up and down in the tubes, with chicken feathers tied to the reeds to create a makeshift air seal. This works something like a conventional bicycle tire pump that uses a plunger system to push the air.

It might be possible to fabricate some type of squirrel cage fan blower using mostly plywood and sheet metal if no functional blowers are available. The basic system is comprised of a series of flat blades or flaps attached to a shaft like a paddle wheel, which is housed within a covering that allows it to fan air into a pipe when the shaft spins. An existing fan, pump, or blower from a broken appliance might also be salvaged to provide the basic air-circulating mechanism, and a makeshift hand-crank system might be adapted from the gears and pedal assembly from a bicycle to turn the paddle wheel. Air can be piped from the blower to the forge using almost any kind of tubing, including steel, copper, or even plastic pipe, with a steel nozzle into the basin (called a tuyere) that directs the air to the hot coals.

Other Blacksmithing Tools

Anvil

Anvils are available in different styles, sizes, weights, shapes, and specific features. An anvil for metalwork is essentially a hard, solid platform upon which hot metal can be shaped by hammer blows. Large, flat rocks have been used as anvils in ancient times and practically any thick block of steel could be used as such. Popular makeshift anvils in modern times have included sections of railroad rail, heavy I-beams, engine blocks, large mandrels, and other odd hunks of scrap steel. Scavenger hunting through an old junkyard might turn up something usable.

The anvil should be set up fairly close to the fire to minimize the heat loss in the piece of metal to be worked. A glowing heat usually only lasts for a matter of seconds in an iron or steel object of typical working size, and then the piece quickly hardens and must be returned to the fire and reheated before continuing. A good blacksmith learns how to work quickly and efficiently to make the most of the fading heat.

A common base for an anvil is a tree stump or large, round log set on end. The surface of the base should be trimmed level. The anvil can be chained or strapped to the base, or large nails or spikes can be driven into the wood around the anvil and then bent over its feet to lock it down. A sturdy stand

Miscellaneous makeshift anvils and forming tools.

Small anvils are usually not expensive, and they are quite useful for numerous small tasks.

such as a small table might also be constructed of wood or steel. The main concern is that the stand is sufficiently solid and stable, capable of supporting the anvil at a comfortable working height.

Most books recommend mounting the anvil where its face, or its working surface, is just high enough that the knuckles of one of the blacksmith's closed fists would touch it when he stands next to the anvil with his arms hanging naturally at his sides. The desired anvil height will be different for blacksmiths of different stature.

Quality factory-made anvils can be difficult to find used in good condition. The best available modern anvils are usually made in the United States and Europe, such as those by Vaughan/Brooks, Mankel, Refflinghaus, Peddinghaus, and a few others, and most of them are not exactly cheap. As I write this, Centaur Forge in Burlington, Wisconsin, sells a range of new anvils and other blacksmith tools and related supplies, as does Pieh Tool Company in Camp Verde, Arizona.

A new high-quality steel blacksmith's anvil of respectable size (say 100 to 300 pounds) will typically cost anywhere from $500 to well over $1,000. Less expensive, new cast iron anvils weighing up to 70 pounds are

routinely sold by Northern Tool + Equipment Company in Burnsville, Minnesota, for under $100. Serviceable older anvils occasionally show up in antique stores and at farm auctions. For an older anvil in serviceable condition, $1–$2 per pound might be considered a fair price range.

Certainly one of the most famous anvil styles used by modern blacksmiths and farriers is the London pattern, with a small cutting table and horn at one end, a square hardie hole to accept hardie shanks, and a round pritchel hole for punching, usually near the square heel in the face. Another popular style of anvil has a horn, or "bick," on each end and no heel.

The anvil horn is a rounded, tapering, and normally pointed extension projecting out horizontally from one or sometimes both ends of the anvil. A horn can be helpful when forming curves and bends in things like rings, chain links, and large hooks.

The anvil's hardie hole can be used to anchor any of a variety of stakes or special hardies having tapered square shanks with the corresponding size. If an anvil lacks a hardie hole, the various forming tools might be set in a block of wood or in a hardie plate made of steel, or the tools' shanks might be

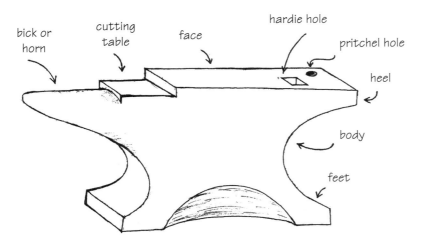

Parts of the anvil.

An improvised cutting hardie made from an old chisel.

ground down or in some cases sleeved to fit the anvil. An assortment of miscellaneous forming tools might also be manufactured from other things like ax heads, trailer hitch balls, wide-bit cold chisels, or other old tools.

Swage

Another potentially useful tool in a blacksmith's shop is a swage block, which contains various holes, grooves, and depressions of different shapes and sizes. The swage block is really an anvil full of odd shapes for special forming tasks. For instance, an oval depression found in some swage blocks is typically used to form spoons and ladles. In the absence of a steel swage block, a spoon might be formed in a depression gouged out of hardwood. I used the wood depression method before I bought a factory-made cast steel swage block, and it works well as a temporary (though smoky) means to a rustic hand-made steel ladle.

Vise

The blacksmith's vise, also called a leg vise, post vise, or sometimes a box vise, is a very handy item. Larger than standard bench vises, it is made to handle heavy hammering. Its jaws close against the resistance of a large spring mounted in the frame, which helps the

The blacksmith's post vise.

 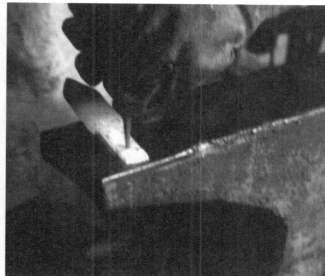

Punching a hole in hot metal over the pritchel hole of the anvil.

device sustain the shock of the hammer blows. It also has an anchor leg or post that extends to the floor for added stability.

Old blacksmith's vises in good condition are not as common as they once were, and newer versions are normally quite expensive. A good one is probably well worth the effort for major forge operations, but a large bench vise is still a lot better than no vise at all.

MANIPULATING METAL

As previously suggested, simple blacksmithing techniques can be applied to alter and manipulate blank material in a number of different ways. Without having a drill press available, holes can still be punched through hot metal. Hot metal can be sheared over a cutoff hardie or cut on the anvil's cutting table using a sharp chisel and hammer, often much faster and easier than by using a hacksaw. Without expensive welding equipment, pieces can still be forge welded together in some cases, or joined by rivets. The blacksmith also has the advantage of being able to transform the shape of the metal without removing any material.

In metal work, upsetting is the process of making the work piece shorter and thicker by hammering on the top end of a bar or cylinder as it stands vertically on the anvil. It is the process applied to rivets to expand them inside the holes of the pieces to be locked together. The opposite of upsetting is drawing out, where the work piece is stretched, normally by peening or hammering, to make it longer or wider.

Whenever cutting or shearing a work piece on a cutoff hardie, the hammer blows during the final stage of the process should be offset to avoid having the hammer hit the cutting edge of the hardie. Practicing this technique will save a lot of time at the grinding wheel reshaping the edges of dulled hardies.

There are quite a few different types of hardies to serve different functions. A cone-shaped hardie, for example, might be useful in shaping rings or forming funnels. A forked hardie is a handy tool for bending bars of steel. Fuller hardies are sometimes used to form grooves. A blacksmith might also create an assortment of unique hardies as he needs them for special tasks.

A drill having either a conventional twist or auger-type bit without any special attachments will cut only a round hole; however, holes of other shapes may be punched through hot metal. Punching normally

Thinning the edge of a spearhead or dagger.

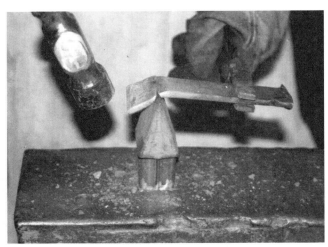

Cutting on the anvil hardie.

displaces the material in order to form a hole, whereas a drill cuts the material out to form the hole. Naturally, there are advantages and disadvantages to both methods.

There are several different means for connecting pieces of metal. With very limited tools, pieces might be bound together with cord or wire or glued together using some kind of strong epoxy like JB Weld. Somewhat stronger connections might be achieved by soldering, brazing, or welding two pieces of metal together, or two or more pieces with aligned holes may be bolted together or securely riveted.

Riveting

The practice of riveting two or more objects together to form a sturdy connection was very common in the old days. The process is really simple, easy to accomplish under most circumstances with basic tools, and normally very effective. The hardest part is probably getting the necessary holes properly positioned in the pieces to be riveted.

Solid, straight rivets can be made from mild steel rod, common nails, fence wire, heavy copper wire, or other similarly shaped pieces of malleable metal. They should be cut to length a little bit longer than the combined thickness of the materials to be riveted. If

pieces of wood, bone, or antler are to be riveted together, washers should be used at both sides where the rivet ends expand to minimize the chances of splitting the material. (I learned this the hard way.)

To rivet two things together, the corresponding holes must first be drilled or punched through the pieces to be joined. The diameter of the holes should be the same or slightly larger than the diameter of the rivet, and it works well if the rivet fits into the holes just tight enough that it can be pushed through by hand or with taps from a mallet.

With the rivet positioned in the holes so that a short length of its ends protrudes equally from the holes on both sides, one end is supported on an anvil surface, while the other is peened down with a hammer. The rivet is upset, becoming shorter and thicker, and tends to form a shape similar to an hourglass. After one end is peened down into the shape of a dome larger in diameter than the hole, the work is flipped over and the other end is peened down the same way. The separate pieces will be tightly locked together as one piece.

This method can be improved by cutting a bevel into the rim of the hole at the openings on the sides where the ends of the rivet protrude, the same way holes are beveled to

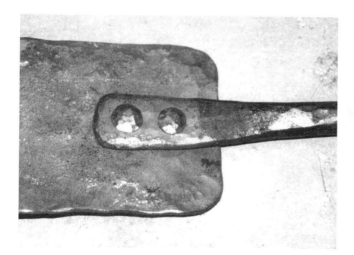

A strong connection using rivets.

This leather knife sheath was riveted together with sections of thick copper wire.

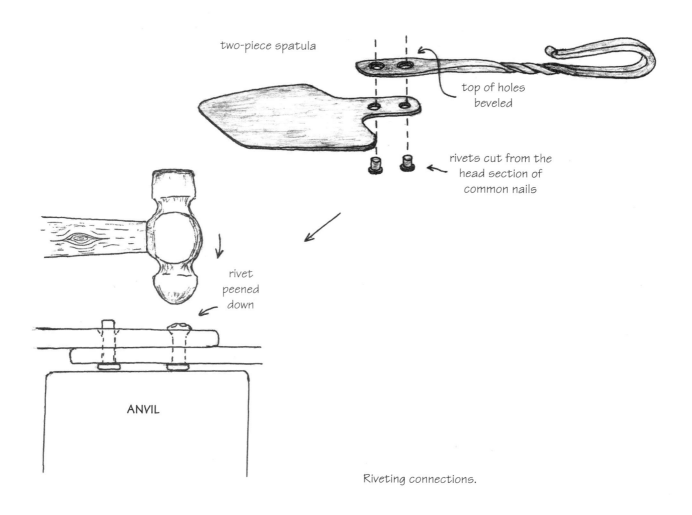

two-piece spatula

top of holes beveled

rivets cut from the head section of common nails

rivet peened down

ANVIL

Riveting connections.

countersink screw heads. This is easily accomplished by drilling into the existing holes using an oversized bit to a prescribed depth just enough to bevel the top of the holes. The resulting funnel-shaped openings allow the mushrooming rivets to swell into those spaces and lock the work together. Using this technique, the riveted pieces can be sanded or ground flat, leaving no raised dome-shaped rivet heads above the surface, resulting in a potentially more refined appearance of the end product.

Rivets made of soft malleable metal, like mild steel or copper, are the easiest to use, and are normally peened cold. Hard steel would have to be heated sufficiently before it could be formed into functional rivets.

Common nails of appropriate diameter make excellent rivets, since they are fairly soft. Also, the heads of the nails can be left on and used for the rivet heads on one side in some situations. With only a limited amount of practice, it is possible to create some very secure permanent riveted connections.

Forge Welding

Welding is a popular method for joining steel. There are several different modern welding methods, including gas, arc, MIG, TIG, and plasma-arc. All of these require somewhat sophisticated equipment. Forge welding, on the other hand, requires only a heat source, hammer, anvil, tongs, and borax or other flux.

There are some tricks to successful forge welding with modern steels, and not all blacksmiths are skilled at it. The process demands that the metal worker pay close attention to certain details. If the forge fire is dirty or the heat range is too high or too low, the results will usually be less than satisfactory. Getting two pieces of semi-molten steel to join permanently with hammer blows can be frustrating business. Even so, it can be done with practice, and has been done with success many, many times before.

Joe DeLaRonde describes forge welding in five simple steps in *The Book of Buckskinning*

IV. His steps are building a clean fire free of clinkers, heating the metal, fluxing the metal, reheating the metal to the welding temperature, and finally removing the metal from the fire and pounding it together.

Clinkers are hard globs of impurities that form at the bottom of a coal fire and tend to hinder a clean, even fire in the forge. When cold enough to solidify, clinkers should be removed from the forge using either a long hook or tongs. It is a sound habit to always clean out the basin of the forge and remove the clinkers before firing it up each time, and then again from time to time as the coal continues burning.

As most of the impurities burn or melt out of the coal, it turns to coke, which has more of the characteristics of charcoal and burns easier and cleaner than fresh coal. Coke is the crusty lightweight carbon material that coal transforms into as it burns, having a texture somewhat similar to Styrofoam. Fresh "green" coal burns with smoke and yellow flame, while coke burns with a blue flame. Coke is what you want for a welding fire.

Most forging activities can be accomplished with the steel in the temperature range where it glows from a cherry red to a bright orange, normally between 1,400°F and 1,750°F. Welding requires higher heat. The proper welding temperature is different for different materials. Wrought iron, for example, can safely reach 2,500°F without burning because it contains almost no carbon.

Carbon steel, on the other hand, must be welded within a narrower range. If heated too hot, the steel will burn and crumble into pieces when struck with a hammer. The proper weld heat for most low carbon steels is somewhere around 2,000°F, but an experienced smith will recognize the right temperature at which the pieces of metal will stick together. This is usually a brightly glowing buttery yellow, just before the showers of sparks begin erupting from the metal.

Scale forms on the surface of steel when it is heated to a light cherry red, at around 1,600°F, and it begins to flake off when the steel reaches an orange glow, above 1,700°F. This scale forms when the hot metal is exposed to oxygen, and it interferes with a good weld. It must be removed before the pieces to be welded can properly stick together.

Most blacksmiths use some type of flux to remove these oxides from the metal surfaces. Typical ingredients used in fluxes for forge welding include iron filings, baking soda, salt, and borax, such as 20 Mule Team Borax. The flux functions like a solvent by combining with the scale and melting it away. It should be introduced to the material to be welded when the scale starts forming prior to reaching the welding heat. I find that a long-handled spoon is handy for this.

The weld has a fairly good chance of taking hold if all the conditions are met: a clean, coke-burning fire is used to slowly heat the metal to the right heat; the pieces are sufficiently fluxed at the right stage; and are quickly removed from the fire when they reach the precise sticky welding temperature; they are placed upon the anvil with one piece directly over the other where the connection should be; and then the pieces are pounded together with one or several hammer blows straight down.

However, a lot of things can ruin the weld. If carbon steel reaches the white-hot temperature where sparklers explode from its surface, it will normally be ruined at that point. If the surfaces of the pieces to be welded aren't free of scale, ashes, or other contaminants, they won't stick together. And if the hammer blows aren't delivered straight down over the pieces, the pieces will skid apart rather than stick together. Also, the blacksmith must work fast when the right temperature is reached, because the metal won't hold the welding heat very long at all after the pieces leave the fire. Successful forge welding demands patience, understanding of the process, and plenty of practice. (Scale, sparks, flux, and globs of semi-molten metal splatter in all directions when the hammer comes down on the pieces being welded. Spectators should stand well outside of the splatter zone during this process.)

A basic understanding of steel's changing characteristics at different temperatures can be especially useful because different objectives are often achievable through various heating and cooling processes. High-carbon steel, for example, can be hardened simply by heating it to a bright cherry red glow (approx. 1,500°F) and then quenching it in either oil or water. In this state the material's crystalline structure has been shocked, so to speak, and is under substantial internal stress, making the material actually somewhat brittle.

Tempering
Some of this stress can be relieved in the process known as "tempering," whereby the work is reheated to a much lower temperature, usually 800°F or less. This might be followed by any rate of cooling. Tempered steel will be softer than it was immediately following the heating and quenching process, but it normally retains good strength, has improved resilience and less crystal distortion, and is less subject to cracking.

Carbon steel that was previously hardened or tempered can be drastically softened by the process known as "annealing." The steel is heated to a temperature above its transformation range, or to a higher temperature than where its crystalline structure changes (more than 1,500°F for most high-carbon steels), and then allowed to cool very slowly, allowing its crystal transformation to completely settle. This relieves the stresses in the material and makes it much easier to cut or machine cold.

Slow cooling is best accomplished by burying the work in hot coals, sand, or ashes and allowing it to gradually cool with the material that surrounds it. Keeping the work covered until it is completely cool prevents exposure to cold air during the process, which might cause rapid cooling much the way quenching does. When properly annealed, something as hard as a file should be softened to the point where it can be drilled, filed, or cut with a hacksaw. After the machining operations are done, the work can again be heat-treated to restore its hardness.

As noted earlier, plain low-carbon mild steel lacks sufficient carbon to allow it to be hardened by heat-treating. However, it can be case hardened. In this technique, the work is packed in carbon-rich material, usually in a crucible, and heated to its transformation range so that its outer layer absorbs enough carbon for it to be hardened on its surface. In this way, the body of the work remains fairly soft while the surface is hardened and possesses increased wear resistance. This can be useful with certain machine parts and tools.

Gas Welding, Soldering, and Brazing

Forge welding and riveting connections were just discussed at length, but there are times when other methods for joining things will be appropriate.

A gas welder, like an oxyacetylene torch, can be used to cut, weld, braze, and solder. This equipment, which will produce a flame of 6,300°F, can be used for lower temperature tasks, such as brazing and soldering, by reducing the oxygen and gas pressures at the regulators. Most soldering and some brazing operations are achievable using propane, butane, or MAPP gas (a liquified petroleum gas mixed with methylacetylene-propadiene). A small torch using either propane or MAPP gas might be ideal for small soldering jobs, as long as a supply of the gas and the solder are available.

In soldering and brazing, the solder or braze metal (usually a rod of some alloy of brass) melts and adheres to both pieces, thereby joining them; the base materials are not actually fused together as with a weld. A fairly solid joint is possible.

The keys to successful soldering and brazing are using flux properly and ensuring that the base metals are sufficiently hot so that the solder or braze metal will flow over the surfaces as needed. Soldering and brazing generally work well where the ends of pieces to be joined overlap or are bridged with a separate piece that splints the two main parts together. The typical MAPP gas torch

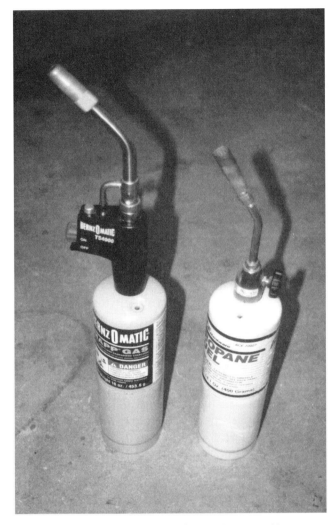

Small, inexpensive gas torches for soldering and brazing.

currently sells for around $40 or less, with spare cylinders of gas normally averaging around $8 each.

TOOLS YOU CAN MAKE

Making Files

Perhaps one of the most useful tools typically employed in the shaping and refinement of certain other tools is the file. Weygers describes how to make files in *The Complete Modern Blacksmith* by raising file teeth on a smooth surface of high-carbon steel using a

A crude Bowie-style knife forged out of a heavy bastard file.

Making Other Tools out of Files

Cutting tools should be especially hard in order to be able to hold an edge. Files are about as hard as tools can be, but they are also comparatively brittle. If you try to bend a cold file, it will not bend; it will eventually snap. Strength, hardness, and toughness are not synonyms. A material might have high tensile strength, but not be particularly hard. Likewise, a material that is considered extremely hard, like a file, is generally brittle, lacking the toughness we would expect to find in a more ductile material; obviously a knife made from a file would be a poor choice as a prying tool. Toughness can be defined as the ability to resist shock, while hardness relates more to wear resistance.

Because of their superior hardness, old worn or broken files usually lend themselves well as stock material for making other cutting tools such as knives and chisels. I have also found that files make excellent flint strikers for producing sparks. A lot of old frontier knives were forged out of files and, although not as tough as spring steel, file steel nevertheless will hold a finer edge than most other available steels.

Thread-Cutting Dies

I have found that functional thread-cutting dies can be homemade from files. Any file thicker than 1/8-inch will work. To make the die, the file is first heated to an orange glow and then allowed to slowly cool. When completely annealed, the file is no longer difficult to cut. It can then be drilled using the appropriately sized drill bit wherever desired, and three "cloverleaf" grooves filed out with a smaller file in three different directions from the center hole, evenly spaced. These cutouts serve as spaces where the displaced metal chips can escape, and also provide cutting edges inside the hole that will ultimately cut threads on a shaft. Careful filing of these grooves should leave sharp enough edges in the hole to facilitate an adequate cutting action.

sharp cold chisel. The material must first be annealed to soften it. Rows of file teeth are created by forcing the sharp edge of the chisel into the surface of the work, thereby raising the teeth by wedge action.

Individual round teeth, such as what you would want on a wood rasp, are made using the same principle, but with a narrow punch rather than with a cold chisel. It is noted that handmade files are comparatively crude and cannot compete with factory-made products. Even so, under special circumstances a set of makeshift files could be very important.

Operational thread-cutting die made from a thick file. This will easily cut threads on a 3/8-inch rod.

The thread-cutting teeth in the die can either be cut cold with a corresponding tap (if such is available), or the file can be heated in the forge and a screw or bolt can be screwed through the hole in the hot file to create the desired cutting teeth. Dimensions change somewhat with the heating and cooling, and this is a crude way to make a die, but it is workable for many applications. The final step is to reheat the file and quench it in water or oil to restore the original hardness.

A threading tap might be similarly fabricated from a round file, such as a rat-tail file, with the filing teeth either ground off or more precisely removed using a metal lathe. The file should first be annealed to make it soft enough to receive the necessary machining operations. Three deep grooves should be cut lengthwise, either by chiseling or engraving, or possibly by using a milling machine or a Dremel Tool, which would serve the same purposes as the cloverleaf grooves in the die. The thread-cutting teeth could be formed using the appropriate die, before the long flutes are cut. The original hardness of the file should be restored by heating and quenching after all of the cutting is done to form the tap, just as with the makeshift die.

Drill Bits

Drill bits can also be made out of files. A tapered, flat spade bit is simple to make from a piece of a small file, using a grinding wheel to sculpt the functional dimensions. A three-cornered file will also make a functional triangular bit, which can be designed to bore holes as well as to function like a reaming tool. But files can't sustain much torque.

Whenever grinding blades or small tools on a fast-spinning grinding wheel, care should be taken to avoid burning the metal. The work should be dipped in water every few seconds to keep it cool, and heavy pressure against the spinning grinding wheel should be avoided. If the edge of the steel starts turning different colors during the grinding, that portion is likely ruined. Also, protective clothing is advised, as grinding wheels have been known to fly apart at a higher rpm. Imagine multiple fragments detaching from your bench grinder at high speed and sailing across your shop like bullets! It's always good to check for cracks in the wheel before turning the motor on, to wear safety goggles, and to go slow and easy with the grinding. Obviously, slower, manually turned grinding wheels would be safer in this regard.

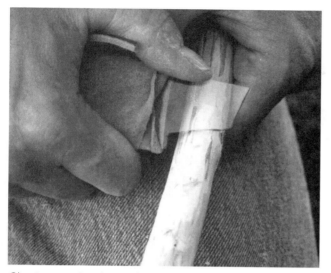

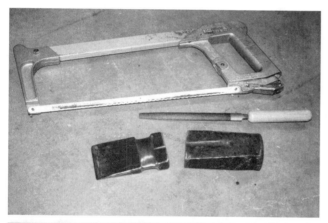

Shaving wood with a knife made from broken glass.

The hatchet shown in the two photos above was made by filing a groove around a log-splitter wedge, sawing off the extra wedge material, and affixing a handle.

Making Knives

Without using a grinding wheel, a fine-cut flat file, or a sharpening stone, steel knife blades might still be honed to a fine edge using a common abrasive rock. Decent knives have been made out of leaf springs, files, large saw blades, lawn mower blades, and other scrap steel having the ability to be hardened. In rare situations where steel of sufficient hardness might be difficult to find, a suitable cutting edge might be found in a piece of broken glass. I have found that a plain edge of broken glass, even without being knapped, usually serves as an excellent scraping and shaving tool. The biggest disadvantage of glass is its fragile nature.

Making Axes and Hatchets

The ax is a versatile tool that is often considered next in importance to a knife, especially in forested regions where felling trees, chopping firewood, and trimming or shaping lumber for numerous purposes are all routine activities. A good ax is also a useful emergency tool for penetrating rooftops, locked doors, or walls where people might be trapped during a flood or fire. Survivors caught without axes or hatchets might be motivated to improvise makeshift versions of the basic tool.

Axes have been made out of a variety of other things. Those who've seen the movie *Cast Away* (2000) will remember the ice skate lashed to a stick as a makeshift ax. I made a functional camp hatchet from a 4-inch section of a log splitter's wedge, and it served the purpose well. I filed a groove around the makeshift ax head to receive the handle, and then wrapped the handle made of split, bendy green saplings around it, finally securing everything together with long strips of rawhide and glue. The steel in one of those wedges is fairly soft, but I used this hatchet to cut firewood and hammer tent stakes into the ground during a remote backpacking trip a few years ago; I am confident that a factory-made hatchet would not have handled these tasks any better.

It is sometimes feasible to heat a bar of steel in the forge and then hammer it into a wedge shape to form an ax head. The eye through which a conventional handle would fit can be either punched or drilled through the thickest part of the head. If the resulting hole is smaller than desired, it can normally be filed into an elongated oval shape.

Tomahawks and belt axes, or hatchets, were common products of frontier blacksmiths. The usual method was to fold a heated, flat bar of wrought iron over a round steel mandrel to form the eye for the handle and then hammer the two iron ends together to form the blade, or bit. Often a narrow piece of carbon steel was sandwiched between the two ends of the softer iron strap as they were forge-welded together to create a bit hard enough to hold a decent edge. The bit would be hammered into a flat, narrow taper, and finally honed to a cutting edge on a grinding wheel.

A comparatively flat bar of steel might also be filed or ground to a sharp edge on one or both ends, then set into a split branch handle and secured into position with either strong cord or rawhide. A flat ax head such as described might also be cross-pinned to its handle. A heavier head would of course make chopping easier and probably hold up better under heavy use than a much lighter flat ax head.

Making Screwdrivers and Punches

Certain tools, like screwdrivers and punches, are quite easily fabricated from short lengths of carbon steel rod or certain types of alloy steel. Mild steel would normally be considered too soft for such purposes and some tool steels might be harder than needed. The material should be hard enough to resist deformation but not so hard as to be brittle, where normal stresses might cause fracture. Flathead screwdriver tips and tapered punches are easily shaped on a grinder.

Making Clamps and Vises

Methods for clamping and securing work are routinely needed when certain sawing,

A small belt ax forged out of a 1-inch diameter square bar. Note that the handle fits into a hole through the head.

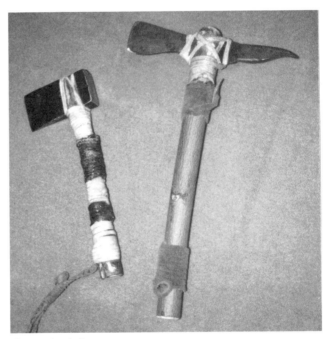
Two makeshift axes.

A sturdy four-way T-handle screwdriver made by brazing two screwdriver tips together.

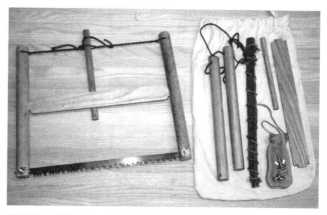

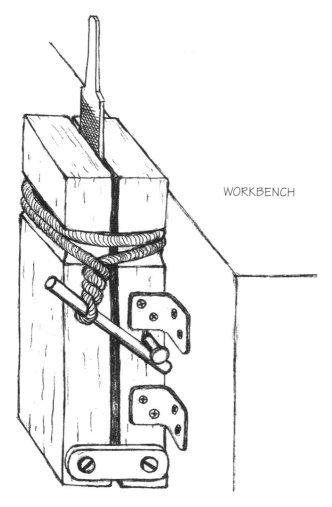

WORKBENCH

A makeshift vise made from 2x4s, with the clamping action provided by a twisted rope.

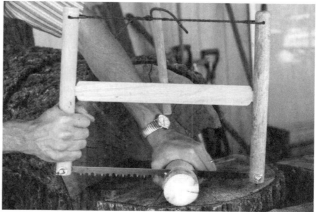

A bucksaw can be easy to assemble and disassemble and uses twisted cord to provide the tension for the blade. Besides the blade, all that's needed to make a bucksaw are sticks, nails, and a leather bootlace.

Simple, primitive clamp using a split branch, pivot stick, strip of rawhide, and wedge to create the clamping action that in this case holds a bone arrowhead firmly in its jaws.

drilling, or filing operations are performed. Human hands are amazingly versatile holding/gripping machines, but they lack the strength required for certain tasks. In situations where steel clamps or vises may not be available to hold objects as needed, makeshift clamps can sometimes be devised.

The clamping power of twisted rope or strong cord is considerable and, when used in conjunction with sturdy jaws, will provide an amazingly tight grip. Hinged jaws of wood or metal can be strapped together with cord and tightened with the tourniquet method, using a steel rod or hardwood dowel slipped under the rope and turned to twist the rope up and tighten the jaws. A large nail or bolt can be set into the base and used as a stop for the dowel to prevent the rope from untwisting. This same system of twisted cord is used in conjunction with two levers that pivot over a center bar in a bucksaw to supply tension to the blade to keep it rigid.

The common system used in modern C-clamps and bench vises employs a screw to tighten the jaws or grip plates. Screw clamps and vises provide superior holding power, and might be duplicated by a competent metal-smith under certain circumstances.

In ancient times, long wooden vise jaws were sometimes loosely banded together with rawhide somewhere near their middle with a thin board or dowel spacer between them serving as a pivot, and a wedge was forced between the jaws at the end opposite the work being held, causing the jaws to pinch the work. The jaws could be opened when needed simply by knocking the wedge loose with a mallet.

Making Manual Drills

Drilling holes without electricity can be accomplished using several different methods. Manually operated drills are still commonly sold in hardware stores. Brace drills and auger bits are simple and sometimes practical machines used for boring holes in trees or lumber. The eggbeater-style hand-cranked drills are sometimes useful wherever electrical power hasn't been hooked up. The old Yankee pump drills and pump screwdrivers that were once popular for small projects still turn up at yard sales on occasion.

In ancient times, two other manual drilling systems were also sometimes employed. The simplest is probably the bow drill, where the cord of a short hand bow is wrapped one turn around a spindle that holds the boring bit. Sawing back and forth with the hand bow turns the spindle and, by applying downward pressure on a socket that fits over the top of the spindle, a hole is easily drilled by the rotating bit.

Another ancient drill machine is the Indian pump drill, which consists of a shaft, or spindle, a hard drill bit, a flat board with a

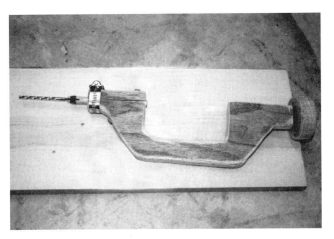

A crude but effective homemade drill brace.

large hole in its center that serves as the pump platform, a suspension cord, and some sort of flywheel. The spindle loosely fits through the large hole in the platform, allowing the platform to slide freely up and down on the spindle. The platform is secured at each end by the suspension cord, and the cord is also supported at its center as it passes through a hole near the top of the spindle. The cord can be quickly adjusted so that the platform is suspended level. Several turns of the spindle will raise the platform as the cord wraps around the spindle.

The drill bit is rotated by the unwinding action of the cord around the spindle as the platform is pushed down, and then again as the cord winds back up, raising the platform. The added weight of the flywheel gives the rotation some inertia, helping the bit cut and also keeping the spindle rotating until the suspension cord winds itself back up where it can be pushed down again for the next spin. This system produces a fairly smooth drilling operation when everything is done properly. The entire tool is also simple to build and, like the bow drill, requires only wood, cord, and some type of hard material such as steel or a sharp stone for the drill bit.

A primitive bit can be set into a slot in the end of the spindle and secured with wrappings of cord, or in some cases cross-pinned. I

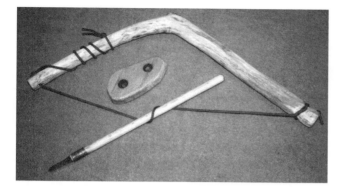

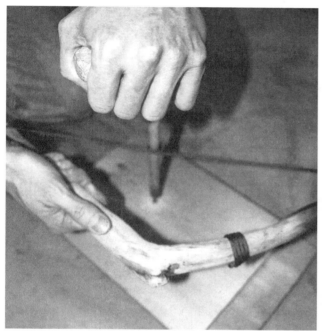

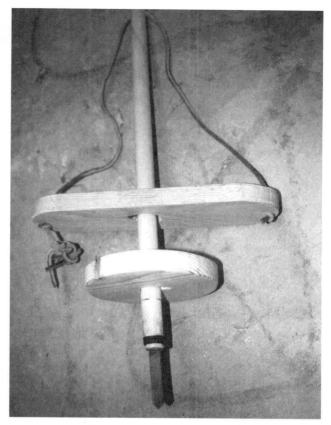

A simple bow drill and how it is used.

Makeshift T-handle attached to a twist drill.

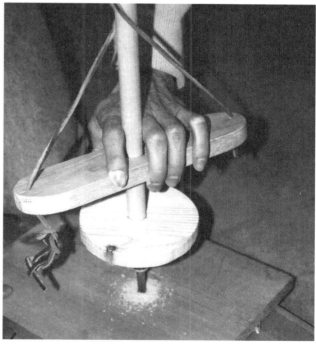

The Indian Pump Drill and how it works. A spindle has a makeshift bit inserted at one end, and it is turned with a platform, flywheel, and cord.

designed my pump-drill spindle to accept removable bit holder tips so that I can change bits as needed for different drilling jobs. The holder tips are secured to the spindle with removable dowel pins.

The simplest manual type of drill is probably some style of auger or twist bit with a T-handle that can be slowly turned by hand to bore holes. Some of the early stone drills were, in fact, shaped like a T and apparently turned by hand in this manner. This type of simple pointed and mostly flat stone drill would bore holes essentially the same way as the modern flat spade bit does.

Making Miscellaneous Hardware

In addition to assorted hand tools, numerous items of hardware would eventually be needed by survivors in a Dark Age, just as miscellaneous hardware is routinely needed by people in stable times. However, in a Dark Age there might not be the convenient well-stocked hardware stores open for business downtown. Survivors might have to make their own nails out of thick steel wire or small springs out of saw blades, for example.

Nails were commonly made on the frontier and can be simple to make. A block with nail-sized holes of the proper gauge prevents the stems from bending while you upset the

ends to make the heads. The pointed ends of the nails are then hammered out into a point on the anvil after the heads are done. The task of shaping the nail is easily accomplished when sections of the metal are heated to an orange glow. Soft steel is the usual material for most types of nails and spikes.

Various bolts and screws are easily made from sections of steel rod, using thread-cutting dies like the homemade versions discussed earlier. Whenever cutting threads, it normally works best to perform the cutting with short, partial turns of the die, and to periodically reverse the direction for a turn or two between the cutting turns, to clean up the back threads and to help keep the die on track. I normally use cutting oil to cool and lubricate the cut and a small brush to sweep chips off the work every few turns.

Hooks, latches, brackets, staples, and washers are among the simplest hardware to make, involving only very basic metalworking techniques. Things like chain links, hinges, and belt buckles can be more intricate with the fitting of multiple parts, or sometimes with welding pieces together. If hinges of metal are unobtainable, I have heard of functional door hinges being made by nailing scraps of heavy cowhide to the door and its frame to serve this purpose, as a temporary

A fire block scraper, small arrow point, sewing needle, small knife, and a short pack saw all made from hacksaw blades.

A saw blade set into an antler handle and secured with rawhide.

expedient. I would expect this to work best with small, lightweight doors, such as on cabinets or crates. But metal door hinges are normally not extremely difficult to build in a metal shop.

STONE TOOLS

For many thousands of years, stone tools were used to complete essential routine chores, from killing and butchering animals for food to processing hides for clothing, or constructing weapons, shelters, and other necessities. I maintain that a basic understanding of stone-tool technology could be beneficial to a survivor—if not as a total substitute for more advanced tools, then certainly as a useful supplement wherever supplies of refined metal stock or other modern-day materials are severely limited.

It seems fitting that this chapter would include some basic stone tool manufacturing techniques, given this technology's place in history. Although primitive, stone tools can still be put to good use for a number of purposes.

Certain rocks possess characteristics that make them well suited for specific purposes. Tough, heavy hammer stones can be used for pounding, sandstones can be used for sanding, and sharp-edged stones can be used for cutting. Rocks have also been used as tools to modify other rocks to create better tools.

A variety of rocks have structural qualities similar to glass and behave similarly in the way they break. Fracture lines spread along somewhat predictable angles to the point of force, forming a shape-altering break, or knap, in the material in the shape of a widening cone. This is known as a conchoidal fracture. As Paul Hellweg points out in his book, *Flintknapping: The Art of Making Stone Tools*, a BB striking a window often creates this kind of cone-shaped break in the glass. A familiarity with this principle allows the knapper to chip or break rocks that possess this characteristic in a predictable manner.

Man-made glass is the perfect material for applying the principles of conchoidal fracture

because it contains few flaws. Obsidian is sometimes referred to as nature's glass and is perhaps the easiest of the natural minerals to knap. Other siliceous rocks include flint, chert, chalcedony, agate, opal, jasper, quartzite, petrified wood, and ignimbrite. Some of these are easier to knap than others, but a rock must be somewhat vitreous (glasslike) to be knapped properly.

Heat-Treating Stone

Archaeological evidence shows that some prehistoric people used heat treatments to alter the crystalline state of their stone, and a lot of modern-day flintknappers use ovens and kilns to bake their raw material before modifying it into implements. Flint and chert are commonly heat-treated. There has been some debate about what technically occurs in the structure of the stone during heat-treating, but the purpose is to make the material easier to knap.

Effective heat-treating is accomplished by subjecting the stone to a gradual increase in temperature until it reaches a critical range between 400°F and 1,000°F (usually around 500°F for most siliceous rocks), keeping the material at this temperature for several hours, and then allowing it to cool very slowly. Stone Age people used fire pits in the ground to thermally treat their stone, using layers of sand or dirt to sandwich their stone blanks between layers of hot coals, which would hold the heat in the pit as long as possible. The process could take up to several days for best results. Rocks to be heat-treated must be dry— I learned many years ago that rocks that contain moisture, as plenty of stones collected from a fresh streambed do, can explode when heated in a campfire. I would compare such an experience to having a high-powered rifle round discharge close by when you least expect it.

Methods of Stone Working

Stones can usually be modified into functional tools using one or a combination of four basic methods: 1) percussion flaking, 2) pressure flaking, 3) pecking, and 4) grinding, or abrading. Percussion and pressure flaking are used to modify flint, obsidian, and other glass-like rocks, while pecking and grinding are typically used on grainier stones that lack the fracture predictability of glass.

Percussion Flaking

Percussion flaking is the technique of removing material from a blank or core by striking blows, either directly with a hammer stone or heavy antler billet, or indirectly using a hammer and usually a punch made from antler.

Pressure Flaking

Pressure flaking is the technique of removing material from a blank by pressing flakes off using a pointed tool, which is often made from an antler tine or from a short section of copper wire set into a wooden handle. With either percussion or pressure flaking, the angle of fracture is what determines the results of the process more than anything else.

Typically, a blank or blade is detached from a main core, or cobble, with a blow from a hammer stone (percussion), and then this blank is worked into a more refined arrowhead or small tool using the pointed antler tine with the pressure flaking method. Individual techniques may vary somewhat. A tiny arrowhead could be produced entirely by someone with the necessary skill using a percussion method. In any case, the conchoidal fracture principle will apply when one of the glassy or flint-like stones is worked.

A right-handed flintknapper would normally hold the work in his left hand and press off flakes with the flaking tool held in his right hand. The left hand, in this case, is best protected with a leather glove or leather pad, as the stone blank and any flakes removed from it can be extremely sharp. Eye protection is also advised. The chips or flakes detach with force and occasionally fly in unpredictable directions. Some knappers also drape a piece of leather or canvas over their laps to keep thin slivers of stone off them-

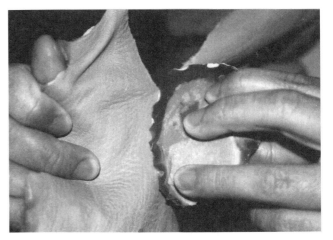
The razor-sharp edge of a flint can cut through tough hide.

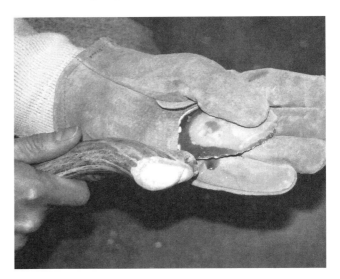
Percussion flaking using an antler hammer.

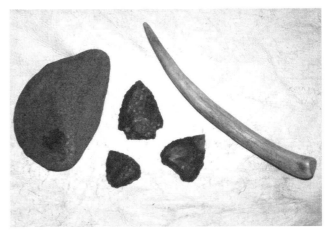
Flint points created using the pressure flaking technique.

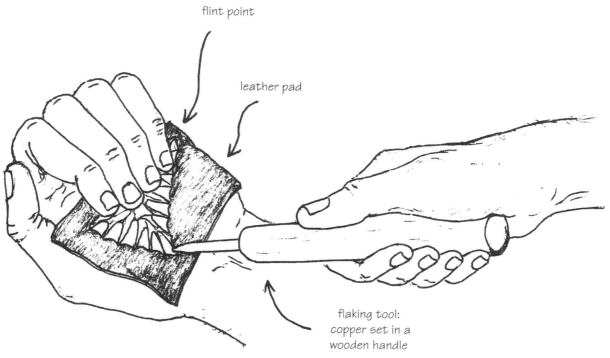

flint point

leather pad

flaking tool:
copper set in a
wooden handle

Pressure flaking.

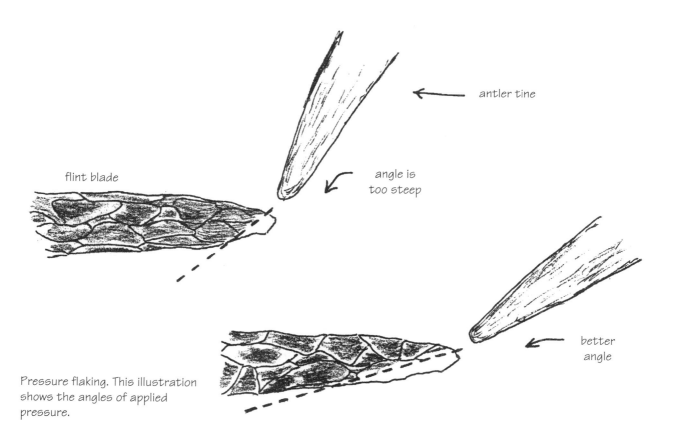

antler tine

angle is
too steep

flint blade

better
angle

Pressure flaking. This illustration
shows the angles of applied
pressure.

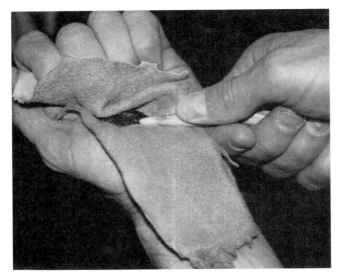

Pressure flaking using an antler tine. Note the protective layer of leather between the sharp tools and the stoneworker's palm.

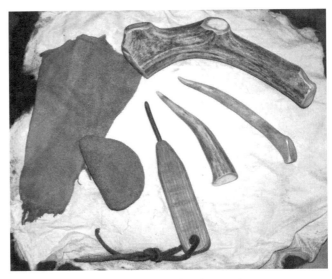

Tools for working stone include leather pads to protect the hands, sandstone for roughing up slick edges, pressure flakers made of deer antler and copper, and antler hammers for percussion flaking.

selves as much as possible. Disposal of waste material should be handled methodically as well, and the stone chips should be treated the same as broken glass.

The pressure flaking technique involves pressing the pointed flaking tool of either copper or antler into the edge of the blank to press off thin pieces of stone and thereby trim down the edge as desired. Materials like deer antlers or copper are suitable for pressure flakers because they are sturdy enough to sustain the necessary applied pressure, yet soft enough to bite onto the edge of the stone without slipping in most cases. A working edge of a blank that is too thin or slick to facilitate proper tool contact can be roughed up or ground down with a sandstone or other abrasive rock. A copper flaking tool can be made easily by inserting a short length of heavy-gauge copper wire (1/8-inch to 1/4-inch in diameter) into the end of a wooden dowel and grinding or filing the tip into a dull point or taper. Some experimentation might be useful, as each individual will find his own preferred tool shapes and dimensions.

A cutting edge is produced by pressing the flaking tool into the edge of the blank at a downward angle. Flakes are removed from the face of the underside of the blank as it is held firmly in the palm of the left hand (for a right-handed knapper), which is enveloped in a protective leather pad. The angle of pressure will affect the length of the flake removed. Pressing almost straight in toward the edge of the blank will normally press off a longer flake, resulting ultimately in a thinner, better-looking edge. Using a sharper downward angle of pressure results in shorter flakes removed, and a thicker edge on the final product. The process usually requires some practice to master. Beginners typically press off flakes that are shorter than desired until they acquire the proper technique.

Arrowheads are commonly notched at the base where wrappings of thread or sinew can best secure them to an arrow shaft. These notches are easily produced using a pressure flaking tool with a slightly finer-than-average point, or with a small notching wrench made from bone, antler, or hardwood.

Grinding

As already mentioned, grainier stones and stones that lack glass-like characteristics can

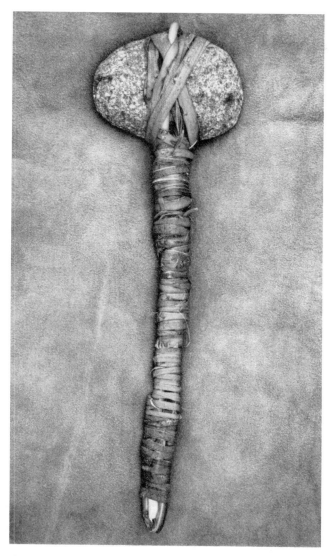

A stone hammer of granite with a handle made of rose stems and bark.

Pecking

Pecking is the process of removing material from a blank by striking its surface repeatedly with a harder hammer stone, thereby crumbling away layers of material where needed to achieve a desired shape. As with grinding, the process of pecking can be tedious. It was a common method for creating hafting grooves around stones. A combination of pecking and grinding might be practical in the manufacture of stone tools for emergency survival.

Common Stone Tools

Properly shaped and securely hafted stone axes and celts have proven themselves when it comes to chopping down trees and splitting logs. Creating a functional handle attachment to a grooved ax head can be achieved in several different ways. Probably the most common was finding a small, pliable, green tree branch and thinning it where it would bend over the ax head, then wrapping the ends together with vines, twisted bark, or rawhide in order to pinch the stone head securely in the bend. A celt is a somewhat rounded stone ax head made with no groove for hafting; typically, celts were wedged into elongated holes through thick branches serving as their handles.

Another very basic type of stone tool is a simple scraper described by Paul Hellweg in Richard Jamison's *Primitive Outdoor Skills*. A handy scraper is created using a technique referred to as "bi-polar percussion," in which a hard, rounded, fist-sized stream cobble is placed upon another rock serving as an anvil and then struck a sharp blow on its top. The process, if executed properly, is likely to shear the cobble into two pieces, leaving a fairly sharp edge where the rock broke apart.

Very effective cutting tools can be made from glass or stone. *Primitive Technology: A Book of Earth Skills*, from the Society of Primitive Technology, includes a letter by Errett Callahan, PhD, that advocates stone arrow points for modern-day hunting. He refers to a 1984 American Medical Association announcement stating that obsidian can be as much as 500 times sharper than surgical steel.

be modified into usable tools with other methods. Granite, for example, was often used for hammer stones and was shaped as needed either by pecking or grinding or a combination of both methods. Harder, finer-grained material like basalt greenstone would be more suitable for axes, where a sturdy tapered edge must be honed for chopping. In either case, the method of using an abrasive rock such as sandstone or coarse-grained granite to grind down or abrade another stone requires more patience than precision.

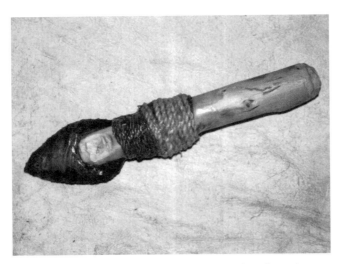

A small flint blade is secured to a wooden handle with sinew and twisted grass, plus a generous amount of glue.

Wooden mallets are easy to make. The example on the right was simply cut from a chair leg.

Using good knapping techniques, somewhat precision stone knives are achievable.

We have now explored various stone weapon points, knives, axes, and hammers. Some of the other tools that have been made of stone include adzes, mortars and pestles, drills, hide scrapers, hand choppers, sanders, grooving chisels, saws, awls, and splitter wedges.

OTHER NONMETALLIC TOOLS

Besides stone, primitive tools were sometimes also made from antler, bone, and hardwood. Knives, awls, and sewing needles were commonly made from animal bones, and horn and antler were similarly used to construct a wide variety of usable implements, from flaking tools and hammers as already described, to digging implements, scrapers, and handles of all sorts.

While neither antler nor bone can ever be as hard or made to be as sharp as the hardest stones, they have the advantage of being generally tougher than stone, or at least less brittle. An antler arrow point, for example, would lack the heavier weight that most good stone points of equal size would provide, and most likely give an inferior performance on game in terms of flesh penetration. But the antler point might be used again and again, being less likely to break than the stone point. Additionally, both antler and bone are usually much easier to drill holes through than any rock harder than sandstone.

So, it can be seen that different natural materials have different advantages that make each suitable for certain applications. The key, as I see it, is to become familiar with the physical properties of all of the natural and man-made materials, and to understand how they can all be used to benefit a person's life. A survivor's quality of life will surely depend on his resourcefulness to a great extent, and this means using whatever can be found to improvise the tools that will be needed.

Bones, horns, seashells, ivory, and antlers can all be worked using abrasive stones. A bone can also be split where desired by first chiseling or scoring a groove around the intended split line using a sharply pointed hard rock such as flint or agate, and then tapping the bone along the groove with a rock to split it into two pieces. Bones are hollow and are usually easier to cut through than a piece of antler of equal diameter.

Naturally, wood can be used to construct almost anything—walking canes, clubs, splitter wedges, tool handles, fence poles, shelter frames, pack frames, gun ramrods, gun stocks, springy bows, arrows, spear shafts, stir sticks, containers, and digging sticks. Again, an abrasive stone might be used to sand the material, and sharp rocks can be used to cut, whittle, or scrape it into usable dimensions.

Hundreds of potentially useful things grow on trees and other plants. Functional fire tongs can usually be made from green branches. Some tree barks and plants contain usable cord fiber or basket-weaving material. The possibilities are too numerous to list.

CHAPTER 4

Recycle and Salvage Everything

Usable containers are thrown away every day. Many could be used for storing supplies or water during a Dark Age.

Most of us would probably be amazed if we knew how much the average American family discards each year. We throw away reusable glass jars, bottles, coffee cans, plastic jugs and containers of all kinds, sturdy cardboard boxes, broken appliances, ugly furniture, worn clothing, scrap metal, and literally millions of other items we consider to be either junk or just too much of a hassle to repair, clean, or find room for. And we're all guilty. Throwing things away is how we clean out our homes. Ours is a consumer society, and if we never disposed of what we couldn't use, we would soon bury ourselves in our own garbage.

A surprising amount of what we send to the landfills is perhaps recyclable, and some organized recycling programs have tried to minimize the impact that a growing population's trash inevitably has on the environment. But as society becomes increasingly accustomed to conveniences of all kinds, the demand for disposable products naturally increases as well.

A coffee-can skillet is used to boil water, heated with a small alcohol stove made from a pop can. (A good method for making one of these is described by Alan Halcon in *Wilderness Way* magazine, Volume 10, Issue #3.)

In a world where commerce and mass production are severely crippled, such as what is contemplated in this book, people would be forced to salvage and recycle existing products to a greater extent. Many kinds of things we would normally discard without thought now, we might consider finding new ways to use in a future Dark Age when most new products could be precious and scarce.

This chapter will explore ideas on how we might make good use of miscellaneous trash to expand our limited resources in a devastated world. We will take a look at how certain products that have exhausted their original usefulness might be modified or adapted to serve other useful purposes or simply repaired and reused as originally intended.

During the Great Depression of the 1930s, and also during WWII, patching and reconditioning well-worn used clothing and other equipment was very common. During the war, certain products and materials were in short supply to civilians due to the demands of the government in the war effort. When new tires weren't available, old tires were recapped. Other products were similarly scarce, and creative methods were employed

by the resourceful civilians to extend the life of existing used goods.

Even during times of prosperity, salvage is big business. There are hundreds, if not thousands, of wrecking yards across the country where used auto parts are stripped from junked vehicles and sold at a fraction of their new prices. During a Dark Age, these places might be the main source for replacement car parts for at least as long as automobiles are still in use. Other products, machines, and appliances would similarly be cannibalized as they wear out or break down, in order to keep others in service when new replacements are no longer available.

If the world becomes a desolate wasteland littered with broken-down vehicles and dilapidated empty buildings after millions of people have been annihilated in a nuclear holocaust or decimated by plagues, jury-rigging skills would clearly be among any survivors' greatest assets. At least for a period of time, people would have to figure out how to use whatever they find to serve their needs.

Survival-minded individuals already tend to contemplate such things regularly. The shiny aluminum in pop cans is occasionally consid-

Broken tools, such as this hand trimmer, can sometimes be made into other tools.

Tons of usable scrap lumber, as well as other used building materials, get hauled off to landfills routinely.

A functional knife blade fashioned from a broken pair of grass clippers. A protective sheath is made from cardboard and electrical tape.

A simple stew-can stove, fired with Sterno.

ered for fishing lures, for example, and bent nails are shown in most survival books as potential fishhooks. Coat hanger wire is commonly valued for its excellent utility, and large, heavy-duty plastic garbage bags are sometimes seen as potential rain ponchos, among other things. Once a person develops a habit of visualizing the things in his environment in terms of how they might be made useful, the possibilities begin to appear endless.

Lumber prices have steadily climbed in recent years, even as truckloads of salvageable used and scrap lumber get hauled to the dumps along with other debris from demolition and construction sites in every part of the country. It is normally considered too labor-intensive and time-consuming to pull out all the nails or screws and clean up the wood for reuse. But when resources are scarce, those debris piles will get picked over carefully for usable wood, we can be sure. And we can expect that most of the pulled nails would be straightened and reused as well.

Even if some of the lumber is split, broken, twisted, or weathered beyond any structural usefulness, it might still serve as firewood for woodstoves and fireplaces to heat homes in the cold winter when split logs might be scarce and natural gas and electricity are no longer available. People might consider themselves fortunate to have anything at all to burn, once the catalogs and phone books have all been consumed.

Roadside scavenger hunting can be a fun and fascinating hobby. Some treasure hunters use metal detectors for the metal relics and coins hidden under the ground, but the amount of interesting items to be found on the surface or in the weeds along most highways is also amazing. If you like to walk (it's good exercise, anyway), tie a canvas sack to your belt and head out to the nearest highway some Sunday morning when the weather is decent, and see what you can find. When the traffic is not too heavy, treasure hunting can also be accomplished from the car to some extent. It always amazes me the kinds of things people routinely lose from their vehicles, especially along rural highways. You just have to be alert as you drive in order to spot the good items lying on the side of the road.

Much of what most people consider trash could be valuable in a real-life survival situation. Odd lengths of rope, string, and wire are common along roadsides. Disposable cigarette lighters litter plenty of roadside curbs and trails. Most of them are found empty, but occasionally a discarded lighter will still contain a small amount of butane, possibly just enough to light a couple of campfires. Even most of the completely empty ones are still good for producing sparks.

Parking lots are another kind of environment worth exploring. I've found a lot of tire weights in parking lots. They are useful for casting bullets or for making fishing sinkers, as they consist largely of lead. Most people will not pick up tire weights when they see them lying on the roadside or in a parking lot, but lead costs money. I have also found coins, interesting bird feathers, and miscellaneous small items of hardware in parking lots, simply by paying attention to what lies in the gravel or on the pavement. A good scavenger hunter learns to spot useful objects among the common trash.

Good things occasionally show up in Dumpsters behind stores and shopping malls. Most stores send damaged products to the trash bins, but a lot of what ends up in the trash is repairable or usable for other purposes.

The activity known as Dumpster diving can be fruitful, but a hopeful treasure hunter should also understand the risks involved. Broken products often come with jagged splinters or dangerously sharp edges. Also, most commercial trash bins are stored on private property and the business owners understandably tend to prefer that scavengers stay out of their trash, mainly for liability reasons. These same risks typically exist around debris piles at construction sites as well. It is usually advisable for the treasure hunter to obtain permission from the owners first, whenever possible.

All lakes and rivers that have been frequented by fishermen, swimmers, duck hunters, or boaters have bottoms littered with man-made objects. A scavenger hunter with a small boat and a strong magnet tied to a long rope can haul in some interesting treasures. People are always losing things in the water. Steel objects such as old tackle boxes, pocketknives, fishing reels, very small outboard motors, and miscellaneous sporting goods are all likely prizes. Rust is the primary issue to be dealt with here.

Magnets can be found in electric motors, which can be found in a lot of appliances, shop tools, pumps, and plenty of other devices. These motors eventually wear out, but their magnets, coils of wire, bearings, screws, and certain other parts are sometimes worth salvaging for other purposes.

Considering that automobiles and large machines contain thousands of individual parts, many of which might be stripped off and used for other things, exploring abandoned junkyards could be a profitable venture in a post-collapse era. Converting salvage into usable products can be an enterprise that is only limited by the imagination.

Previously we considered certain tools we could build from other products. I remain convinced that most of what we need to get by in life could be derived from many of the routinely discarded objects we normally view

These prizes were collected while scavenger hunting.

Broken hardwood chairs provide good material for tool handles or even firewood, if nothing else

as articles of worthless rubbish. Creative minds can devise ways to use a lot of the junk.

The people of most of the world's primitive societies were skilled at utilizing natural resources. Nearly every part of hunted animals was utilized—the meat was eaten, and the skins were tanned and used for clothing, bags, and sometimes for shelters. Brains were used in the tanning of hides. Sinew was used for bowstrings, for thread, and for backing wooden bows. Rawhide was used for binding things together, and scraps of hides were boiled for glue. Antlers, horns, and bones were shaped into containers, awls, sewing needles, and other tools. Similarly, the rocks, earth and clay, trees, and plants found in the environment were used in ways that made peoples' lives better.

Nature is clearly the greatest recycler. The plant kingdom consumes carbon dioxide and produces oxygen, while the animal kingdom consumes the oxygen produced by the plants and produces the carbon dioxide that the plants need. Certain living organisms become food for other living organisms, which eventually die and decompose in the Earth to fertilize the soil and feed the plants that become food for animals that in turn become food for other animals. Nature's cycle repeats itself over and over and over again.

An efficiently operated small farm might take advantage of certain recycling processes. Chickens and livestock feed on grain and hay that might be grown on the farm, and the animals' droppings might be used to help fertilize the soil for the crops.

Backyard vegetable gardens and fruit trees are currently not uncommon in the residential neighborhoods of many American cities, and I imagine that these food-bearing plants would be even more popular in a Dark Age when the availability of groceries might be considerably limited. This probably depends a lot on the availability of water for the plants, however, which may also be in short supply in the drier climates.

We can be sure there would always be resourceful survivors composting organic matter to improve the nutrients of their soil. Banana peels, melon rinds, coffee grounds, grass cuttings, wilted lettuce leaves, and other normally discarded organic waste would be saved and recycled as fertilizer, just as is commonly done now in conjunction with a lot of hobby farms and backyard vegetable gardens. Recycling precious water might require a bit more creativity. Receptacles built to collect rainwater and various filtration systems may become commonplace in some regions.

Homes constructed primarily of hay bales, automobile tires, empty jars or bottles, or various other unconventional building materials became something of a trend in the southwestern United States a few years ago, and I find some of the ideas quite interesting. Old tires stacked up and packed with dirt form the walls of some of the dwellings, often referred to as "Earth Ships," in an effort to extend the usefulness of the essentially non-biodegradable rubber tires. Advocates of the Earth Ship homes note their excellent insulation qualities as well as their structural soundness.

Wood stoves and fireplaces might someday be the primary heat sources for homes in cold regions. Those who burn wood in their homes have to either cut, haul, and split their own firewood or buy their wood from someone else. Most of us throw away a lot of things we could instead burn in our stoves. When we trim back the branches of the trees in our yards, we could save the branches for kindling. When we periodically replace the old weathered or split boards in the fences around our properties, we could save the junk pieces for the wood stove. We might need to scrape off old paint, but wood is wood, and it all burns. (However, specially treated rot-resistant wood is best not burned in homes because of the chemicals in the wood.) A lot of the burnable things we would normally disassemble or rake up and stuff into our trash barrels, like tree bark, pinecones, pine needles, twigs, crate boards, pallet boards, broken wooden furniture, picture frames,

A candle made from spare wax, with cotton rope for a wick.

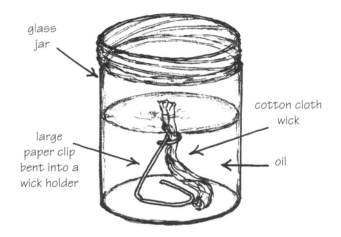

A simple vegetable oil lamp.

Lighting with the vegetable oil lamp.

Coffee cans are really quite versatile.

cracked baseball bats, broken broom handles, phonebooks, and even newspapers, could be potential fireplace fuel. Where fireplace logs are in short supply, people might burn tightly rolled newspapers banded with wire as a substitute. Even cow dung has served this purpose in barren regions.

Convenient lighting in the dark is usually provided by electric lights in our present time, but makeshift lanterns might replace them when the power is out, or when batteries and lightbulbs are in short supply.

You can make an excellent long-burning lamp by filling a small glass jar halfway with vegetable oil and adding a wick made from a half-inch wide strip of canvas or other cotton cloth. Bend a large paperclip to hold the wick.

I made such a lamp using a small empty peanut butter jar, and after 12 hours of continuous burning, the jar still contained more than half of the original oil. This type of lamp or lantern burns longer than most of the candles I have used, and I believe it is generally safer than a candle. If the oil lamp were to tip over, the flame would most likely self-extinguish in the spilling oil. The vegetable oil should burn well even if it is old or rancid and no longer suitable as a food product. The grease and fat left over from frying can also be used in makeshift lamps.

My mother has always tried to curb wasteful habits at my parents' house. Any food left over from a meal is almost always saved for other meals. Dishes are routinely washed with the minimum amount of water. Freezer bags used to keep foods other than meats or cheeses are usually cleaned out and used again. A lot of the extra fruit from the trees was canned and saved for later, rather than being allowed to spoil. Several years ago she made fresh soap out of the tiny scraps left over from other used soap bars.

Something as common and disposable as an empty coffee can might effectively serve a number of useful purposes. A section of coat hanger wire can be affixed to the can as a bail to turn it into a small easy-to-carry pail. A folded strip from another metal can might also be bent and soldered to the side as a handle. A small camp stove can be fabricated out of a metal coffee can, with holes for ventilation cut with a knife. The more common function—as a decent storage container—makes it difficult for me to throw empty coffee cans away.

Much like automobiles or other machines, firearms consist of numerous individual parts, and they are sometimes lost, broken, badly rusted, or worn out, rendering the whole gun unserviceable in the most severe cases.

Salvaged gun parts are the merchandise of some lucrative businesses.

Keeping firearms operational in a future Dark Age might be especially challenging for some survivors. Weapons may be subjected to harsher than usual weather conditions, proper gun-cleaning habits might be very difficult to maintain for long periods, and replacement gun parts may become increasingly scarce, and eventually as sought after as gold. At the same time, the need for working weapons during desperate times would likely be very high.

It is not uncommon to find complete and working used firearms of all kinds that were built entirely from parts from other guns. In some instances, special fitting is performed by gunsmiths to ensure proper functioning of the different parts, but cannibalizing damaged guns for their serviceable parts is a practical way to extend the usefulness of products that no longer function on their own. Just about any broken or worn-out mechanical device could be restored from salvaged parts in this way, and it establishes one of several important advantages to standardization with vehicles and weapons used by military or survival organizations—to ensure parts interchangeability.

I believe the possibilities for constructing useful things out of throwaway junk is really unlimited. I've seen sandals and moccasin soles made from the treads of used automobile tires. The treads are cut to foot size and then glued to the bottoms of shoes or moccasins using Barge cement or other strong rubbery adhesive, or ropes are attached directly to the rubber to create sandals. Wherever thick leather or rawhide might be in short supply, almost unlimited footgear might be available in the form of old tires.

Clothing should probably never be thrown away, unless it is somehow permanently contaminated. When articles of kids' clothing are outgrown, they can normally be handed down to younger kids and used again. Scraps cut from worn out clothing can often be used to patch holes or tears in other clothing. Other scraps might be used as shop rags, dust rags, wash clothes, potholders, handkerchiefs,

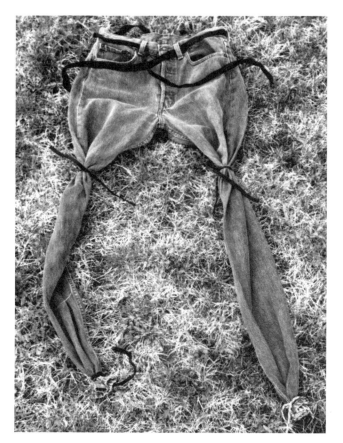

With a few shoelaces or a bit of rope, a pair of jeans can become a backpack, permanently or temporarily.

muzzleloader bullet patches, or gun cleaning rags. Small bags and pouches might be made out of trouser legs, shirtsleeves, or old socks. Thin cotton cloth might be used to strain or filter water. Several garments sewn together might serve as a makeshift blanket.

An old pair of blue jeans is easily converted into a convenient small backpack. Just tie off the legs with shoelaces to form shoulder straps, then connect the ends of the legs to the belt loops or to a rope or belt that closes the top of the pants. It makes a functional bag for carrying other gear on the back, and after you reach your destination, the laces can be removed and the jeans can again be used as pants.

The ability of survivors to creatively use and reuse whatever resources are found in their environment, to include the typically overlooked junk as well as the more obvious natural and man-made resources, will surely have a huge bearing on their quality of life. This may in fact apply as much in our own society with the current trends as they are, as it would after some future global disaster. When we train our minds to ponder the possibilities, we soon discover that they seem to have no end.

CHAPTER 5

Making Fire

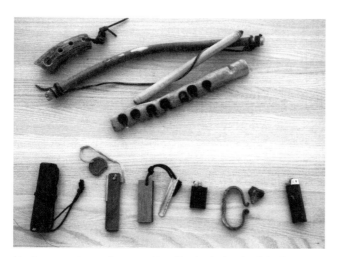

Various systems for creating fire include the friction bow-drill, several different sparking devices, and a cigarette lighter.

Let us now shift our attention to the subject of fire, as I believe this will always be of great importance to our lives in one way or another. In primitive times, fire was used for keeping warm, cooking food, signaling, providing light, hardening clay and wooden tools, and for smoking meat and hides to help preserve them, among other things. In a future Dark Age, I expect that fire might be useful for keeping warm, cooking food, boiling water to purify it, sterilizing surgical instruments, firing clay, heating, tempering, and melting metals, and for providing light wherever electricity might be unreliable or in short supply. I am convinced that the ability to create and control fire will give knowledgeable survivors a significant advantage in the future, just as it has in the past.

As important as fire obviously is to the quality of human existence, it is also potentially very dangerous to the environment and a real threat to the survival of those who find themselves in the path of a raging wildfire.

Attentive supervision and proper safety measures therefore will be every bit as important as the ability to create and effectively use fire. The most common hazards associated with fire are the blaze spreading beyond the hearth or fire pit due to popping embers or wind-blown ashes; soot and smoke filling the lungs; and carbon monoxide poisoning and asphyxiation in confined spaces where the fire consumes all the oxygen.

The majority of problems can be prevented by ensuring that shelters have adequate ventilation; fireplace chimneys are clean and in good condition; campfires or candles are not positioned too close to low-hanging tree branches, dry leaves or grass on the ground, curtains, or other unintended flammable matter; breaks or walls are used to shield wind around the fire pit; proper equipment such as tongs, pokers, shovels, buckets, and thick leather gloves are on hand for safely handling the fire; a supply of water or loose dirt is available to put the fire out; and no fire is ever left unattended.

There are a number of methods by which ignition might be initiated. The chemical reaction we know as fire is really a rather complex process, but we can create and effectively use it for our purposes if we only understand a few basic essentials.

Three things are required in order to make fire: oxygen, fuel, and heat. These requirements comprise what is sometimes referred to as the "fire triangle." Without all three, a fire will not burn. Also, there are two different types of combustion: flaming combustion and glowing combustion. Flaming combustion burns fuel gases, while glowing combustion burns fuel solids. Typically, a burning piece of wood will undergo the flaming combustion process first, until the gases contained in the wood have been completely consumed by the flames, and then the wood will usually undergo the glowing combustion process for an extended period. The glowing combustion is seen as the glowing coals or remaining embers after the flames have died down.

PRIMITIVE METHODS FOR MAKING FIRE

Early methods for making fire, such as the bow drill, the hand drill, or the fire plow, relied on friction to generate the necessary heat.

Fire Piston

One primitive method used in some parts of the world is the fire piston, which utilizes rapidly compressed air to produce enough heat to ignite the tinder. These fire sets consist of a cylinder of hardwood plugged at one end and bored inside to about a half-inch diameter, with a perfectly fitted wooden shaft or piston having a cupped tip to hold a small pinch of tinder, and a push knob on the opposite end.

With the piston started in the cylinder, a sharp blow on the knob with the palm of the hand forces the piston swiftly in, creating the compression that ignites the tinder. The piston is then quickly removed and the tinder is transferred to a larger tinder bundle where it can be fanned to flame. The principle is the same used in diesel engines. While seemingly simple in concept, everything has to be just right for the fire piston to work. In the demonstration I watched, a gasket of thread was wrapped around the end of the piston to create an airtight seal. (To be honest, my own limited experimentation with the fire piston has yet to yield a glowing coal as of this writing.)

Bow Drill

The more common bow drill method for obtaining fire is one that demands a bit of practice to master, but a determined individual with some fundamental knowledge can create fire using very basic natural or makeshift equipment. The fire set consists of a fireboard, spindle, top socket, and a short, loosely strung hand bow. The cord used with the bow must be fairly durable to sustain the friction generated while spinning the spindle.

I personally applied a great deal of effort over a period of time attempting to create fire using the bow drill, quite unsuccessfully at

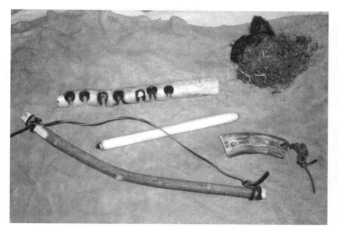

The five main components of the bow-drill fire kit: the bow, the spindle, the fireboard, the top socket, and the tinder.

Starting a hole in the fireboard with the point of a knife. Note the wedge-shaped notch where charred powder can escape.

first but nevertheless producing my share of smoke, until I watched the technique demonstrated in a video and finally learned how to do it correctly.

Different instructional sources recommend different types of wood for the component parts. Some prefer softwoods, while others recommend using hardwoods. My guess is that most types of wood will produce a coal when the proper techniques are applied. I've made it work with pine and cottonwood, though some say that pine is too resinous for best results.

The spindle should be a straight shaft between about 5/8- to 1-inch in diameter, somewhere between about 8–14 inches long, and tapered at both ends like a pencil. The wooden fireboard, or hearth, should be fairly flat on the bottom for stability, and long enough to accommodate multiple drill holes. The drill holes can be started with the tip of a knife blade. Shallow funnel-shaped depressions fairly close to one edge are sufficient to start with. Wedge-shaped notches should then be cut from the closest edge of the fireboard to the centers of these holes, which will allow the charred powder produced by the friction of the spinning spindle to escape and accumulate.

The top socket can be made from a piece of hardwood, bone, antler, or rock that will fit

the palm of the hand. The socket has a depression into which the top of the tapered spindle will fit to stabilize it during the spinning. The depression in the socket should be as smooth as possible to minimize the friction at that point.

The hand bow can be a simple bent stick, to which the cord that serves as the bowstring is secured at both ends. The bow can be a rigid L-shaped stick, about a foot and a half long. The cord should not be strung taut, but with some slack, so that it can be wrapped one time around the spindle with just enough tension to rotate the spindle. The cord will have to sustain a fair amount of friction. Synthetic cord of at least 1/8-inch diameter, such as a bootlace, clothesline, or parachute cord usually works well. Leather, rawhide, and plant fiber cords have all been used in the past. The disadvantage of rawhide is that it stretches. The disadvantage of plant fiber cords is that they tend not to wear as well. Survivors might have to make do with whatever they can find or make.

The key to the bow drill is the coal that must ultimately appear in the generated charred powder. The process will never create fire until the coal is produced. With adequate materials and good technique, a coal should develop within a minute or two of rapid spindle spinning.

The coal won't be visible at first, but its existence will be apparent when the smoke issuing from the charred powder is thick and continues rising from the powder on its own without further friction. By then the powder will be a very dark brown or black, and it will contain a tiny coal that won't be visible until air is gently blown on it.

Very often when people are first learning the bow-drill technique, they will encounter little trouble generating a substantial amount of smoke, but they might lack the acquired intuition for accurately reading the smoke, or fail to understand exactly how to proceed once they have the hidden coal in the charred powder. If the powder keeps smoking steadily after the action stops, it's time to set the spindle aside and nurture that coal buried within. This is the most delicate part of the whole process.

Before starting the spindle to spin, an adequate tinder bundle should be prepared into which the coal can be transferred after it develops. A mass of fine dry fibers, like a bird's nest, makes the ideal tinder bundle. Mosses, crumpled dry leaves, certain fuzzy tree barks, and cattail fluff are all popular for tinder. A flat, rigid leaf or flat piece of bark might be useful under the fire hole in the fireboard to catch the charred powder. A thin, pointed twig is also useful for prying the accumulated powder out of the wedge-shaped notch in the fireboard and onto the flat plate of bark, where it can be gently blown on before being transferred to the tinder bundle, or directly into the tinder bundle. These things should be made ready before the drilling starts. It is also important to have the fire pit ready with dry twigs stacked up to take advantage of the flames that will hopefully soon arise from the bird's nest and have a supply of firewood to sustain the fire once it gets going. Survivors are usually people who learn to think ahead.

The basic technique with the bow drill might be described as follows:

1) Place the fireboard on level ground with one of the fire holes over a flat piece of bark to catch the powder, or bridge the fireboard over a shallow depression in the ground filled with some tinder for catching the powder.

2) Kneel down next to the fireboard on one knee and step on one end of the fireboard with the other foot to stabilize it.

3) Wrap the bowstring once around the spindle's middle section and, holding the components with one hand, fit the bottom point of the spindle into one of the holes in the fireboard.

4) With the top socket in the palm of the hand, fit it over the top of the spindle while steadying the spindle upright. Moderate downward pressure should be applied to the socket and gradually increased when the spinning starts.

5) With the opposite hand holding the bow close to its end nearest the body, a sawing motion back and forth should get the spindle spinning smoothly. The speed can be increased gradually to build up the friction in the fire hole, until heavy smoke indicates that a coal is in the powder.

6) Once it becomes apparent that a coal is in the powder, the spindle can then be set aside and the powder carefully transferred to the bird's nest tinder bundle using the pointed twig, and gently fanned or blown on. The coal will first appear in the charred powder as a tiny glowing red dot the size of a pinhead when the air hits it. As it is fanned or blown on, it will grow. The powder containing the coal should be placed into the center of the tinder bundle where it can spread and flame up. The tinder should be gradually closed around it while a steady stream of air is supplied to it. The smoke will continue to increase until the flames erupt.

If the spindle starts binding or squeaking during the spinning, something isn't right. The components should be closely inspected and any needed corrections made before continuing. Also, if the bowstring stretches and gets too loose to effectively spin the

Demonstrating the basic technique with the bow drill.

spindle, it should be tightened up sufficiently. A small amount of slack in the cord can often be taken up simply by cupping several fingers of the bow hand over it at that end and drawing it into the curve of the bow to tighten it up, holding it taut during the spinning.

Hand Drill

A simpler friction fire method, but one that some consider more difficult for the beginner to master, is the hand drill. The principle of using a wooden shaft to create fiction by rotation in a depression in the fireboard is exactly the same. But with the hand drill, the smaller diameter shaft (normally 3/8-inch to 1/2-inch diameter) is rolled back and forth between the open palms of both hands to produce rotation. The hands spin the shaft one direction and then the other, while simultaneously applying downward pressure. The motion simulates rubbing the palms of the hands together, and they work their way down to the base of the shaft very quickly, forcing the frequent changing of hand position back to the top of the shaft, temporarily disrupting the rotation.

Some books show a variation to this technique that includes the use of cord thumb loops tied to the top of the shaft, whereby the hands can continue rotating the shaft with suspended thumbs, allowing enhanced down-

ward pressure and sustained motion without the problem of the hands working their way down to the base. This simple adaptation makes the hand drill perhaps even more practical overall than the bow drill, once all of the particulars have been worked out.

Flint and Steel

While the friction fire methods produce a smoking coal, the more modern flint and steel fire methods produce sparks, which are actually very small hot shavings of carbon steel. I have never been able to ignite organic tinder directly with a spark from flint and steel, but when the sparks fall on black charred cloth, the sparks will create glowing coals that can then be fanned and introduced to the tinder bundle to obtain flames.

Charred cotton cloth or canvas can be produced easily using virtually the same process as described to make charcoal, but instead of wood chunks, a metal container is filled with pieces of cloth. The lid is locked securely on a container that has only a small ventilation hole to allow gases to escape and, exactly as we do while making charcoal, the container is placed in a fire and cooked until flames emit from the ventilation hole. The limited oxygen within the container permits the charring process, but prevents the conversion of the cloth into ashes.

Flint, agate, chert, and quartz are the most common rocks for creating sparks with steel. I believe the best steel for sparking is hardened, plain high-carbon steel. Most files make fine flint strikers when a long scraping surface is ground smooth on a flat side or edge.

My preferred method for striking flint and steel together entails covering the fingers of my left hand with a leather pad to protect them from the sharp edges of the flint and, holding the steel in my left hand and the flint with my right, striking the sharp flint against the smooth surface of the steel with downward glancing blows over a piece of black charred cloth, directing the sparks onto it until one continues glowing after it hits the cloth. Some people prefer striking the steel against the flint. My method is simply an attempt to simulate the action of a flintlock gun, which I know from experience works very well. In any case, a good tinder bundle must be kept ready to take advantage of a hot coal.

There are more modern variations of the flint and steel system sold in outdoor outfitter stores. They use the same type of flint substitute material (primarily a composition of iron and cerium) used in some cigarette lighters and, when scraped by an edge of steel, this produces extremely hot sparks. The popular flint strikers embedded into magnesium blocks are wonderful products for the backpacker's kit. They throw showers of super hot sparks, and small amounts of the magnesium can be scraped off of the block and into a pile of shavings using a knife blade, quickly providing effective tinder that will possibly ignite even damp leaves or pine needles.

FIREMAKING TIPS

Wet Weather

Getting a fire started in wet weather can be a monumental challenge, even with a healthy supply of matches. A few simple tricks may improve the odds. If survivors are fortunate enough to have road flares, getting an emergency campfire started

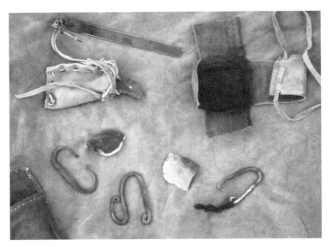

Various styles of steel strikers for the flint and steel kit, with charred cloth to catch the sparks.

Striking a piece of flint against a steel striker (middle photo) and directing the sparks onto the charred cloth to light the fire. A coal is transferred to the tinder "bird's nest," where it is blown into flame (bottom).

HOLE IN
CROSS SECTION

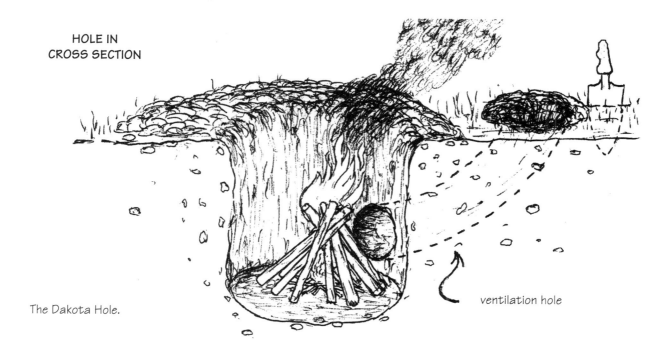

The Dakota Hole.

ventilation hole

should obviously be no problem at all, but we can't always count on having these when a fire is needed.

Standing deadwood is always preferred over damp, rotting trees and branches collected off the ground, or green live wood, as these can be difficult to ignite. Thick branches can be feathered with a series of knife cuts to expose layers of dry material and create finer fuel with better air exposure to aid combustion. Pitch or resin (tree sap) can be the fire maker's savior. It does not absorb moisture like porous wood does, and it is very flammable, even in the dampest of conditions. It burns hot, and will sustain flames usually long enough to dry out other moist tinder materials in adverse weather.

Tinder

Quite often, the difference between success and failure in attempting to create fire will be determined by the tinder. Materials that don't easily ignite can really waste your efforts. Most of the fine, fibrous organic materials make suitable tinder as long as they are dry. Prepackaged tinder or other fuels can be very important to a survival kit. Cotton balls can be rubbed in petroleum jelly to make them more flammable and then sealed in plastic containers to keep moisture out during storage. When they are needed, their fibers can be pulled apart slightly to enhance the air circulation. Fine 0000-steel wool is another good tinder to pack in the survival kit. The wool will burn when showered by the hot sparks from the striker in the magnesium block.

A lot of survival books describe the technique of starting a fire with a magnifying lens. When we were kids we would burn holes in leaves and newspapers with a magnifying glass for fun. This method requires certain elements for it to be successful: sunshine, a magnifying lens with sufficient magnification power, the skill needed to concentrate the sun's rays into a small focal point and then hold it steady until combustion is obtained, and proper tinder or flammable material to take advantage of the generated heat. Burning holes in newspapers on a hot summer day is one thing—creating and sustaining a campfire in adverse weather conditions is something else altogether.

Hiding a Fire

There may be situations when a fire is needed, but building one is risky because the light or smoke may reveal your position to others who are hunting you or who might be considered hostile. The famous "Dakota Hole" is one way to have the warmth of a small fire while keeping the thermal signature to a minimum. It is simply a pit in the ground about the same depth and diameter as a 5-gallon pail that will contain a very small fire. A ventilation tunnel is bored into the base of this main fire hole, angled down from a surface opening several feet away and ideally about the diameter of a pop can, to carry the needed air to the fire in the bottom of the pit.

By keeping the flames below ground level in the Dakota Hole, and using the driest avail-

able wood to minimize smoke, a degree of warmth can be enjoyed and limited cooking can be accomplished, possibly without drawing unwanted attention to the campsite. Such a hole is also easy to fill in when it's time to pack up and move, burying the ashes and erasing the signs of a campfire. If the position of the fire pit is near the base of a tall tree, the branches might serve to help screen any dissipating smoke, minimizing a detectable smoke signature. However, extra caution should accompany the increased fire hazard of a fire close to a live tree, or under any low-hanging branches.

CHAPTER 6

Making and Using Cordage

A surprisingly strong cord braided with twisted grass. Not all varieties of grass are suitable for cord making, however.

The utility value of cordage was discussed in the first chapter, and here the reader will learn how to manufacture functional cord from natural materials. Cord as small as sewing thread or as large as thick rope could be produced by someone with the knowledge, raw material, and patience needed to make it.

CHOOSING THE RIGHT MATERIAL

The first order of business is selecting the right material. Virtually any pliable material that can be bent sharply without breaking can be twisted or braided into functional cord. There are perhaps hundreds of possible different types of natural cord fibers in the world. A few of the best known vegetable fibers found in North America include stinging nettle, milkweed, dogbane (Indian hemp), flax, fireweed, cannabis (marijuana), yucca, willow bark, cedar bark, juniper bark, agave, sagebrush, and cattail. I found a rather unusual variety of long grass growing in my garden with which I was able to produce some surprisingly durable cord. A few other common cord materials include cotton, silk, palm, bamboo, iris, raffia, and jute. You might experiment with what you find in your area.

In addition to plant fibers, animal sinews (tendons), rawhide, tanned leather, intestines (gut), and horsehairs have all been used to construct functional cords as well. There should normally be no problem finding suitable materials with which to manufacture cord in most regions. In some cases, certain synthetic materials might even be available and used in the making of essential survival cord.

Some of the strongest natural cord ever produced by hand is constructed of animal sinew, which has been successfully twisted into bowstrings, snare cords, and tough sewing threads for thousands of years.

Sinew, or tendons, can be taken from either the back of the animal's leg or from along either side of the backbone of deer, elk, or other large animals. This is the tough, stringy tissue that connects the muscles to the bones. The dense white fibers are contained within a leathery outer casing. The sinew is easy to separate from its casing when it is dry by lightly pounding on it with a smooth, round rock or ball peen hammer against a hard surface until it splits open, exposing the usable fibers inside. Care should be taken in breaking open the casing so as not to damage the fibers.

Fresh sinew has natural glue in it that helps hold twisted fibers together, making it especially easy to twist into cord. Like rawhide, sinew also stretches when wet, and a completed sinew cord is best coated with wax or resin to seal it from moisture.

We might turn the raw material into cord by any one or a combination of three different methods:

1) Cutting a narrow strip out of leather or rawhide to create a lace
2) Twisting individual strands or groups of strands of fibers together to form a complete cord
3) Braiding individual strands together to form a complete cord

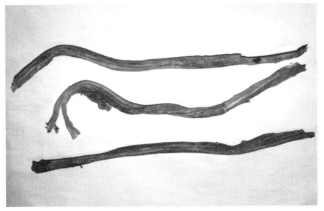

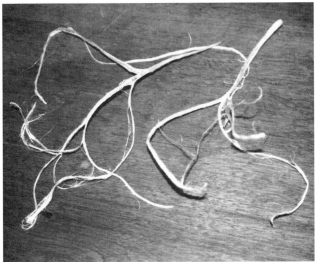

Casings of cartilage from a deer's leg (top photo) contain sinew fibers. Below are the separated fibers.

A lace is cut from a circle of leather in a spiraling progression to the center.

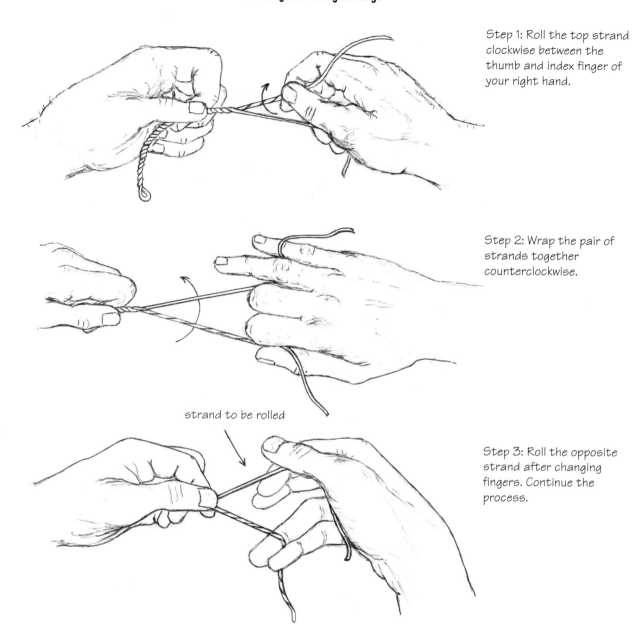

Step 1: Roll the top strand clockwise between the thumb and index finger of your right hand.

Step 2: Wrap the pair of strands together counterclockwise.

strand to be rolled

Step 3: Roll the opposite strand after changing fingers. Continue the process.

MAKING CORD

Spiral Cutting

Using a sharp cutting instrument like good scissors or even a piece of sharp glass or obsidian, a circle can be cut out of an animal hide. A cut can be made into the edge of this circular piece of hide that continues separating a strip to whatever width is desired, following the inner border in a spiraling progression to the center. A surprisingly long lace can be created in this way, and it is about the simplest method I can think of for producing a usable cord.

Two-Ply Twisting

The two-ply twist technique I use to twist cord fibers together is fairly simple, although

THE ELBOW SPLICE IN TWO-PLY CORD.

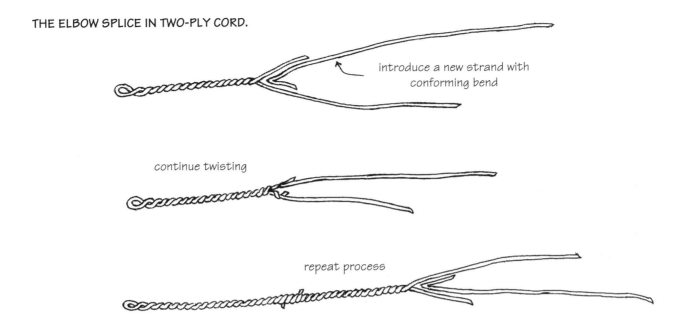

introduce a new strand with
conforming bend

continue twisting

repeat process

it might be considered time-consuming compared to some cord-making methods. I start with a single strand folded over unevenly so that I am working with two running ends of unequal length. I want the two ends offset, as I will be splicing in a new strand each time the shorter one gets close to its end.

The concept is simple. The pair of strands is wrapped one direction while the individual strands are twisted or rolled in the opposite direction, so that the tendency of the fibers to untwist or unravel is hindered by the binding forces of the opposite twist. The fibers are more or less locked together into a long cord.

I find it relatively easy to twist short strands into a long cord by hand. With the technique I use, my left hand serves as a vise to hold the twisted part. With my right hand, I first roll one of the running ends clockwise between my thumb and first finger and, with the other running end pinched between my third and fourth fingers and held taut, I flip both strands over to wrap them together counterclockwise before I switch fingers and roll the other strand clockwise, and then again wrap them together counterclockwise, and so on. The fingers of the left hand periodically work their way up the twisted cord, maintain-

Beginning a two-ply twisted cord with strands of deer sinew.

ing a firm grip close to the spot where the strands are twisted together.

When the twisting process approaches the end of the shorter strand, a new strand is added, overlapping the ending strand for a short length and twisted in with it. Now the previously longer strand is the shorter strand, until a new piece is spliced in to extend that running end, and so on. The sequence can be continued indefinitely, or until there are no more new short strands available with which to continue the cord.

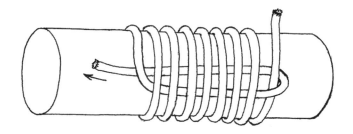

whipping the end of a rope to prevent fraying

binding an eye to a fishing pole

Securing the ends of a cord wrapping a shaft.

Three-, four-, and five-plait braids.

A neat splicing trick that I saw demonstrated in a video by Jim Riggs, the well-known primitive living skills instructor, entails bending a sharp kink in one end of the strand being spliced in. The wedge-shaped fold in the new strand is placed into the V of the juncture where the two running strands are twisted together, so that portions of the new strand are twisted in on both running strands, creating a much stronger splice. This clever "elbow splice" method may create a bulge in the cord where it is applied as a result of the doubling of cord being twisted, but it provides vastly greater holding power over the simple straight splice. Its bulky nature might be countered to some degree by splicing in with the typically narrower tapered leading portion of certain fibrous strands.

A very strong cord can be produced using the two-ply twist technique as described. One of the keys to producing good two-ply cord is uniformity. The two strands that are twisted together should be of comparable diameter, materials having similar physical properties, and twisted using the most consistent technique that is workable. Also, I am convinced that a fairly tight twist produces a stronger cord than one comprised of loosely twisted fibers.

I found that this method of creating cord became very easy with only a limited amount of practice. Indoors, where good natural cord fibers may not be close at hand to practice with, sections of paper towel sliced into narrow strips will serve as a fair substitute, capable of being twisted into a surprisingly functional cord with more tensile strength than one might expect. The technique is exactly the same with any material one might use.

Braiding

Braiding is another method by which a long cord might be constructed from shorter strands. Most young girls are familiar with braids, as the process is commonly applied to long hair for style. A simple three-plait braid is very easy to create and, while perhaps not as secure as a twisted cord of equal material and mass, will produce a stronger cord than any one of the three cords with which it is constructed. The braided cord has a flatter shape than the twisted cord, making it generally more suitable for things like shoulder straps. Three separate two-ply twisted cords could be braided together to form one stronger, larger cord.

To begin braiding three strands, they must first be tied together or somehow secured at one end. As the three running ends lie parallel side by side, one of the two outer strands is first passed over the middle strand and becomes the new middle strand, and then the opposite outer strand is passed over that strand to become the new middle strand. The original middle strand is now the next outer strand in the sequence to be passed over the previous middle strand, and so on. The overlapping strands lock together in alternating fashion. The process is very simple, and much easier to visualize and accomplish than it is to explain.

A four-plait braid will be naturally stronger than a three-plait or three-cord braid. However, the greater the number of cords being braided together, the more intricate the braiding sequence. As with the twisted cord, new strands can be spliced into any braid to continue the process, and running ends of different lengths make splicing manageable.

KNOTS

Using cord of any kind demands some knowledge about knots. Certain knots will be used in sewing, others in fishing, and still others in climbing, sailing, camping, horse handling, and numerous other activities.

Knots can be categorized by their function. Stopper knots are used to prevent rope from slipping through an eye. Bends are used to join two ropes together. Hitches are used for securing ropes directly to other objects like posts, hooks, rings, and rails. Loops are useful to hold the line fast when dropping one end of a line over an object. Shortenings are used to make rope shorter without cutting it, and fishing knots are most often used for attaching line to hooks. Learning how to tie at least 10 of the most common knots is a worthy short-term goal, in my view. One's repertoire of useful knots can be increased over a period of time. The best way to remember how to tie and use certain knots is to practice them over and over again. The first 10 knots on my priority list for mastering follow:

Overhand Knot

The simplest knot is probably the overhand knot, which is frequently used as a stopper knot. It is easily constructed by looping the working end over the standing part and then passing the working end, or running end, back under and through the loop.

overhand knot

Bowline

The bowline is perhaps one of the most important knots a person could ever learn to tie, as it has numerous possible applications. For example, it is a good knot to use for hoisting someone out of a deep hole with a strong rope.

To make the bowline, first loop the working end over the standing part to form a small loop some distance from the rope's end and, by thinking of the running end as a squirrel, the following steps are easy to remember: The squirrel comes up out of his hole, runs around the pole, and then goes back down into his hole.

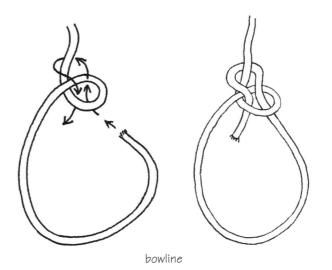

bowline

Square Knot

The square knot (also called a reef knot) is used to join two ropes of equal diameter. It is easily untied when no longer needed by pulling one of the free ends to capsize it. A properly made square knot has a symmetrical appearance, unlike the inferior granny knot, or the very similar looking thief knot, which has its free ends on opposite sides. To make a square knot, remember: "Right over left, left over right."

Sheet Bend

The sheet bend is a very secure knot used for joining two ropes and, unlike a lot of other bends, the sheet bend can be used successfully to join two lines of different diameter.

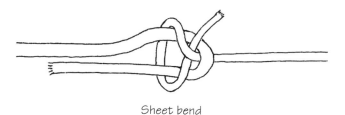

Sheet bend

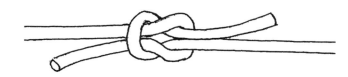

square, or reef knot

Clove Hitch

The clove hitch is certainly one of the most popular hitches. This is probably due to the fact that it doesn't jam under strain and is easy to tie and untie. It works best for hitching rope to round posts as a temporary mooring knot.

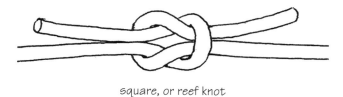

thief knot
(Running ends are on opposite sides. It looks very much like a square knot, but it will not hold.)

granny knot (will not hold)

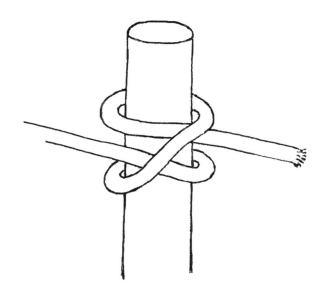

Clove hitch

There are subtle differences between a good square knot and inferior knots such as thief and granny knots.

Prusik Knot

The Prusik knot is an interesting device that has been in use for more than 70 years. It is extremely simple to construct, and it is especially useful for attaching climbers' footholds of medium-diameter rope to a thicker main rope. The knot can be moved around on the main rope when not under load. This knot is easy to form by making a cow hitch and then passing the loop through the knot a second time.

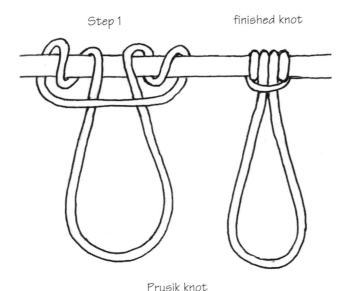

Prusik knot

Figure-Eight Loop

The figure-eight loop has to be one of the easiest loops to tie, and probably one of the strongest. It is bulky and not particularly easy to untie, but it is secure. As its name suggests, it is formed simply by passing the loop through a figure eight and then drawing the knot tight.

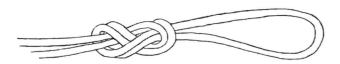

Figure-eight loop

Fisherman's Knot

The fisherman's knot (sometimes called a water knot) is a very good device for joining two small diameter lines of equal thickness. This knot is comprised of two simple overhand knots, one in each line being joined, which jam together for a fairly secure connection. It works much better with fishing line than with rope.

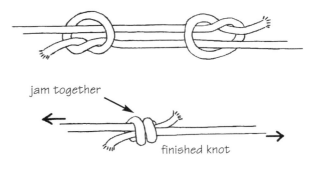

Fisherman's knot

Improved Clinch Knot

The improved clinch knot is a popular knot for attaching fishhooks to line. It works reasonably well with hooks that have eyes and light monofilament fishing line. When properly formed using thin line, it draws up into a tidy knot.

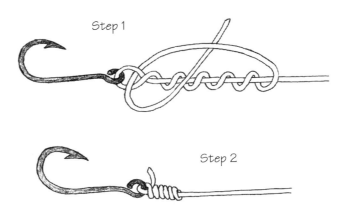

Improved clinch knot

Sheepshank

The sheepshank is a common device used for quickly shortening a section of rope. It will hold together only under tension. The knot grips and holds its shape well when under an equal load at both ends, and is easy to undo when there is slack in the rope. It is also sometimes used to reinforce a weak spot in a rope under tension.

There are certainly numerous other useful knots that might be beneficial to learn and memorize, and I believe that a person could only gain by increasing his repertoire, but quite a lot of common tasks can be accomplished with just the ones described here. One other handy knot that should probably also be included here is the timber hitch, which could be used to attach a bowstring to the nocks of a bow. It's as simple as they get, but quite strong.

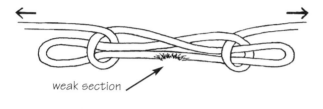

weak section

sheepshank

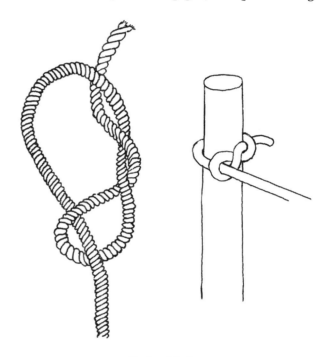

timber hitch

CHAPTER 7

Making Clothing

Shirt and pants sewn by hand using scrap material.

Because the proper clothing will provide our bodies with a degree of protection against the elements, the subject of making our own clothes deserves serious consideration.

In a situation where factory-made new clothing might be scarce, survivors would have to repair and make do with their existing garments, and possibly eventually look toward producing their own from scratch. Wherever supplies of cloth fabric might be found, and where skilled seamstresses or tailors have access to operational sewing machines, this may not present any major difficulties, at least for some time. If the state of civilization continued to deteriorate, however, clothing production could eventually become more of a challenge.

Fashion clothing would likely never be particularly important to survivors, but durable, comfortable, and highly functional season-appropriate clothing will always be valued. In years past, good clothing made from a wide variety of materials—from animal skins to fine silk—has been sewn by hand. Survivors may eventually be forced to learn how to assemble their own articles of clothing, using whatever materials they can get their hands on.

People clothed themselves in animal skins thousands of years ago, and this remains a viable option for certain applications even today. Tanned leather has certain qualities that make it the preferred material for things like knife sheaths, gun holsters, work boots, moccasins, waist belts, hats, horse saddles, handbags, raincoats, wallets, and possibly dozens of other products having utility value.[1]

TANNING ANIMAL HIDES

The five basic methods of dressing, or tanning, animal skins are chrome tanning, acid tanning, alum tanning, vegetable tanning, and brain tanning. The first three methods are usually quicker and less labor-intensive than the older, more natural methods, but the necessary chemicals may become difficult to obtain under survival circumstances. Sometimes a combination of several tanning methods might be applied to one hide to produce leather with certain desirable qualities.

I'll address the other methods of tanning in this section, but the one I'll describe in detail is brain tanning. Although requiring substantial manual labor, primitive tanning methods can produce quality leather. Brain tanning can be broken into eight main steps:

1) Skinning: Taking the hide off the animal
2) Fleshing: Removing meat, fat, and membrane, usually using a dull scraper
3) Soaking: Immersing the skin in a pot or a tub, or possibly weighted down in a stream, to loosen hair and membrane
4) Graining: This is what tanners call hair removal, often done with the hide dried in a frame, using a sharp scraper.
5) Membraning: Removing membrane tissue from the flesh side of the hide, typically using the same scraping tools and tech-

niques used in graining
6) Tanning: Adding the brains, oils, or other tanning formula
7) Softening: Working (and sometimes soaking) the hide until it becomes pliable by keeping the fibers moving during the drying process
8) Smoking: Finally, coating the fibers with smoke, which preserves the hide.

Skinning

Care should always be taken when skinning any animal to avoid nicking the hide with a blade more than absolutely necessary. Every small nick or hole in the hide is likely to tear into a larger hole during tanning. With some animals it is possible (and preferable) to actually peel much of the hide off the carcass with fingers, using the knife mainly to split the hide down the animal's belly and inside the legs, and to detach the hide at the feet and neck.

Fleshing

The first step with any tanning method is to remove the unwanted matter from the skin or hide. The remnants of attached meat, fat, and membranes are typically scraped off the hide using some type of scraping tool. A dull drawknife is popular, used in conjunction

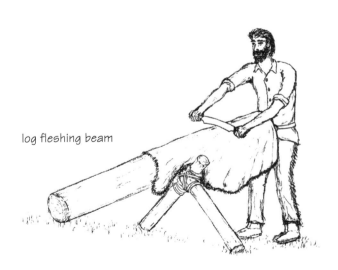

log fleshing beam

The tanner uses a dull drawknife to scrape the flesh side of the hide.

Various hide-scraping blades.

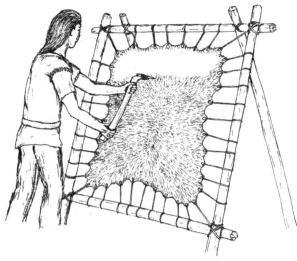

Scraping the hide.

with a round fleshing beam like a smooth log. This step is sometimes done with the moist hide freshly removed from the carcass. Before removing the membranes, some tanners prefer to soak the hide for several days, which allows everything to soften up. Others prefer to lace the hide into an open frame and allow it to dry hard before scraping the flesh side.

Remember: When handling freshly skinned, bloody animal hides, you should wear latex gloves whenever possible. It is messy business, and in some seasons the hair might be crawling with ticks or other tiny parasites.

Soaking

Hair removal is usually the next step (except when tanning fur skins). Soaking hides to loosen hair is common. Ragnar Benson describes soaking hides in a bath with 1 1/2 pounds of unslaked lime for every 10 gallons of water, or substituting hardwood ashes for the lime, with 1 gallon of ashes for every 20 gallons of water. Hides are soaked from two to eight days until the hair begins to slip and then they are removed and scraped clean.

Graining

A lot of buckskinners prefer the dry-

scrape method for removing hair. The wet hide is laced taut into the center of a simple upright framework of poles or boards using strong cord to tether the hide from evenly spaced holes poked through its outer edge all the way around to the framework. The hide is allowed to dry in the frame before scraping. A very large, heavy hide might be more easily staked down flat on the ground for scraping.

Membraning

The best tool for scraping the rawhide clean is usually custom made by the tanner himself and will normally consist of a handle of hardwood, metal, bone, or antler, having a short, curved blade of hard steel solidly secured or set into the handle at a right angle, forming an adze-shaped scraping tool. A 3-inch long section broken off a wide, flat bastard file with one end ground into a cutting edge on a radius works especially well. The curved radius of the edge is needed to prevent snagging and tearing the hide, as would occur with the corners of square blades.

Blades for hide scrapers might also be made from large, flat springs, ax heads, skinning knives, plane blades, lawn mower blades, and possibly even steel buttplates off old rifles or muskets. The harder the steel, the better it will hold an edge. A very sharp edge works best for dry-scraping hides. A sharpening stone will be needed periodically to maintain the sharpest possible edge. The American Indians used stone scrapers before the Europeans brought iron tools into the New World. I once scraped a beaver pelt using sharp, curved pieces of broken glass that I had collected from the trashy ground around a Dumpster; it took several pieces of glass to finish the job because the edges would dull, and I had no good way to sharpen them without creating an undesirable serrated knapped edge. Pieces of broken glass having sharp edges with the necessary curved profile are not always easy to find and by themselves are not as convenient to use as a scraper with a handle, but glass will shave the hair and membranes off a dried animal skin.

The usual scraping technique entails pressing the tool's blade into the hair or fur at a right angle to it as the stiff, dry hide is suspended vertically in the frame, and dragging the scraper to shave the hair off in swaths, using mostly downward strokes with the tool. Each swath should partially overlap the preceding one for thorough hair removal. Also, special care should be taken when scraping around any holes in the hide that could catch the blade and tear the hide. The flesh side should also be scraped in this manner while the hide is in the frame to thoroughly remove whatever remains of the membrane.

Tanning

Brain tanning is among the most primitive methods for preserving hides and furs. It is a type of oil tanning, considered by some professional tanners as not a true tanning method in the strictest sense. Very functional and comfortable shirts, pants, moccasins, and coats have been made out of brain-tanned buckskins. Good brain-tan techniques will produce very soft and elastic skins. Skins are made soft by the very fine oils in the brains of the animal, which lubricate the hide fibers. Finally, these oils are prevented from being rinsed away by water with a smoking process that coats the tiny fibers with pitch, essentially waterproofing them.

The American Indians discovered many centuries ago that the very fine natural oils in brains could be applied to skins to lubricate their fibers. Brained hides were ultimately a lot softer than untreated hides. The brain of an animal contains sufficient oil normally for the same animal's hide. The brains of any large animal can be used. A common ratio is one pound of brains for one deerskin. Possible substitutes for brain oil mentioned in some of the books listed earlier include neat's-foot oil, 10W motor oil, Ivory soap, and egg yolks.

Raw brains could be mashed into a paste and simply rubbed into the skin, or they can first be boiled and stirred into a soup with the hide. The brain oils are absorbed by the hide as it soaks in the soup in a pot or bucket. Some tanners recommend against handling raw brains with bare hands, generally preferring to boil them as a safeguard against possible harmful bacteria. (And you *definitely* do not want to come in contact with the brain or spinal cord if the animal has a chance of being infected with chronic wasting disease, which has been found in deer and elk in some parts of the country.) If the hide being tanned is salted or soaked in a salt-brine, it should be thoroughly rinsed several times to remove all traces of the salt before adding any oils.

In the informative pamphlet *Brain Tan Buckskin*, author John McPherson explains that one need not worry about the hide absorbing too much oil; often the skin suffers from too little oil.

Tanning skins with the hair or fur on is more challenging than with the hair off, because the hair side doesn't dry as easily, and softening the skin without causing the hair to slip can be tricky. Long soakings would not be practical with fur skins.

Softening

After the initial soaking in the brain soup—which need only be a few minutes, just long enough for the hide to become thoroughly saturated—the hide should be wrung out and worked (repeatedly stretched and manipulated) until dry, keeping the fibers moving on themselves. Ideally the hide would then be immersed again in the brain soup for repeated oil absorption before being worked dry again, and so on. This sequence should be repeated several times. The skin will soak up the oils like a sponge. If the skin failed to absorb enough oils, it could easily crack or break wherever folded when dry. Sufficient lubrication of the fibers is the key to pliability and softness.

Smoking

After using the tanning method of your choice, the final step is smoking the hide. If we stopped the process after completing the oiling stage, we would have a soft enough skin that might look good for a while. However, if it ever got wet, some of the oils would likely rinse out with the water, leaving many of the

smudge fire

Smoking hides on a tipi frame.

fibers unlubricated. Eventually the skin would lose its softness when the fibers started bonding together, and it would also be unprotected against bacteria and organisms that feed on dead skin. When a hide is subjected to thick smoke, the fibers are coated with a creosote-like tar or pitch that helps lock the oils in, as well as helping to keep bacteria out to some degree, hindering the decaying processes of the dead skin.

Variations of basic hide-smoking methods have been used with success, but the main objective is to expose each side of the skin to thick wood smoke for about 10–20 minutes per side while preventing the skin from being subjected to the fire's heat.

A commonly employed method of smoking hides involves a series of stovepipes that funnel the smoke from a fire in a barrel or wood stove under a rack draped with hides. Sometimes the hides are affixed to a tripod or tipi framework over a shallow pit in the ground containing a smoky smudge fire. Dead wood rotting on the ground and old crumbly tree stumps will usually provide excellent punk wood that produces thick smoke. It is imperative during the smoking that the hides are kept away from any flames or heat, which could easily ruin them.

Once the hides have been adequately smoked on both sides, the tanning is complete. The leather or buckskins are then ready to be made into clothing.

Other Tanning Methods

Chrome Tanning

Chrome tanning uses chrome salts or crystals, also called chromium sulfate or chromium potassium sulfate. Chrome tanned leather will have a bluish tint. It is well suited for thick hides, like cowhide, or heavy skins with the hair or fur removed. Chrome-tanned leather holds its shape well, is soft, resists stretching, and is virtually unaffected by water. The simplest chrome bath described in Kathy Kellogg's *Home Tanning and Leathercraft Simplified* is prepared with 15 pounds of chrome crystals, 6 pounds of non-iodized salt, and 12 gallons of water. Several different chrome tan recipes are used commercially. Hides are immersed in the solution until completely permeated with the bluish tint, and then removed from the bath and usually rinsed with clean water.

For use with all immersion tans, plastic or wooden barrels are preferred over metal containers, as the salts and other chemicals tend to react with and corrode the metal.

Acid Tanning

Acid tanning, or "tawing," normally uses sulfuric (battery) acid, and can be used in an immersion or a paste. Sulfuric acid is dangerous to handle, so rubber gloves should be worn. The typical acid-salt bath recipe consists of 1 ounce of pure sulfuric acid with 1 pound of non-iodized salt for every gallon of warm water. James Churchill provides a recipe for a three-hour quick tan for a small skin in *The Complete Book of Tanning Skins and Furs* that calls for 3 cups of salt, 2 ounces of saltpeter (potassium nitrate), 1 ounce of borax, 1 1/2 pints of automotive battery sulfuric acid, and 1 gallon of warm water.

Alum Tanning

Alum tanning, also called "dressing," can be accomplished with aluminum sulfate, ammonium alum, or potash alum. Alum tanning is popular for fur skins, as this method actually helps hold the fur in the skin. In some cases a paste might be preferred over immersion when the fur is left on, because extended soaking tends to loosen the fur. Many of these recipes can be made into a paste by adding flour or bran to the mixture.

Vegetable Tanning

Vegetable tanning, also known as "bark tanning," uses tannin (tannic acid) to preserve the hide fibers. This can be extracted from the bark of oak, sumac, hemlock, walnut, and elderberry, from tea leaves, or from various other sources. Ragnar Benson describes using ground and leached acorns. Vegetable tanning is a very old tanning method, and it produces some of the best leather. If you fold a wet strip of vegetable tanned leather, it will tend to retain the folded position rather than snap back into its original flat shape the way chrome-tanned leather tends to do. Vegetable tanned leather is the easiest to mold, shape, and tool. The biggest drawback to vegetable tanning is that it is time-consuming. A large hide might need to soak in the tannin bath for several months to complete the tanning process.

RAWHIDE

Rawhide is skin that has not been tanned. It is comparatively hard when dry, as its fibers are mostly glued together with the natural glue in the hide. Rawhide has its own suitable applications, but when we wish to convert it into leather, we have to soften it.

Soaking the skin in either water or in some special solution will dissolve the glue and make the hide pliable. If we then stretch it, bend it, twist it, press it, wring it, and work it continuously until it dries, the fibers will not have a chance to completely bond themselves back together, and the resulting article will be more pliable than it was in the stiff rawhide state, although it would technically still be rawhide.

SEWING

Sewing the tanned animal skins, or any other available materials, into clothing requires a certain understanding of basic sewing techniques. Assembling garments using a working sewing machine, with plenty of quality fabric and the skills needed to complete the task might be a simple matter, but sewing by hand can be tedious and time-consuming, and awkward for someone lacking any real experience in making clothing.

I know firsthand because several years ago I planned a wilderness excursion, and I was determined to manufacture all of my own personal gear—even the clothing. I had virtually no prior experience making clothing. I scrounged for the needed materials and,

Sandals are useful when camping in hot weather or for crossing streams, as they protect the bottoms of the feet from sharp rocks. I used 10-ounce cowhide for the soles.

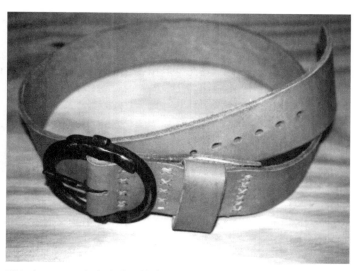

This homemade belt buckle's connections were riveted together.

lacking regular access to a sewing machine, ended up stitching every seam by hand. Over a period of several months I had produced a long sleeve shirt, long pants, wool poncho, wool mittens, two pairs of moccasins, sandals, and a leather belt with a steel buckle. This endeavor involved a lot of trial and error, but every item I created ultimately proved perfectly functional, and it was certainly a valuable learning exercise.

(The fact that it took me a long time and much trial and error to create my wardrobe should serve as a reminder of the importance of specialization. If you can develop a skill— whether it's tailoring, blacksmithing, hide-tanning, or any number of other survival skills—you will have valuable goods or services to offer in barter with other survivors.)

BASIC SEWING TECHNIQUES

Hand sewing is slow and demands patience, but it seems a lot simpler to me than weaving, knitting, or other crafts that build on patterns or require one to keep track of numbered rows or intricate weaves. Connecting pieces of cloth with stitches is about as simple as a task can be,

as far as I can tell. Knowing just two or three common stitches constitutes most of what one needs to know in order to assemble functional clothing. The rest is mainly just measuring, marking, cutting the material, and planning how the pieces will fit together.

Common Stitches

Perhaps the three most common stitches used in hand sewing are the running stitch, loop stitch, and blanket stitch. The running stitch, or "in-and-out stitch" as it has been called, is probably the most commonly used. The loop stitch (also called the whip stitch) is often used with leather lacing. It has the appearance of a spiral notebook binding. The blanket stitch is really a variation of the loop stitch, with the needle passing under each previous stitch before looping over the seam and through the material to form each new stitch. There are more elaborate stitches that might serve special purposes, but a lot of clothing can be made using only the three mentioned here.

The ends of the thread can be secured at the start and at the finish using stopper knots, such as the simple overhand knot that will prevent the stitches from pulling out, or better yet, by passing the needle a few times under the nearest stitch and then through the resulting loop before pulling it tight to cinch it down. With knots it is often possible (and preferable) to sandwich them between seams to hide and protect them from excessive wear.

In my sewing projects I try to keep stitches as evenly spaced as I can, with the stitch holes usually spaced from about a 1/10-inch to 1/8-inch apart. Leather crafters often use overstitch wheels or pronged lacing chisels to mark evenly spaced stitch holes in the leather, using them sometimes in conjunction with a stitching groover that will allow stitches to be recessed and aligned with a consistent spacing from an edge. My own sewing jobs have been somewhat less refined, as I simply judge the spacing and stitch lines by sight. I discovered that practice with hand sewing tends to improve the results noticeably.

A lot of American Indians and frontiersmen preferred fringed buckskins and fringed wool robes to plain garments. The idea was that the fringes would, to some degree, wick rainwater off the clothing. I don't know whether or not this is scientifically sound, but it seems reasonable. The concept might be tested by dragging a dry leather bootlace through a small puddle of water to see how much water the leather would draw up out of the puddle. In any case, fringes are easy to make just by cutting any oversized edges protruding from seams into thin strips. In the end, the goal is that every item of clothing made will be completely functional and practical, comfortable, and as durable and long lasting as possible.

Choosing Thread

Cotton threads tend to hold their stitches and knots a bit better than slippery synthetic threads in my experience. However, they also rot faster and in many examples are not as strong. Linen thread is usually very strong and holds knots well. I prefer sturdy waxed linen thread for most of my leather

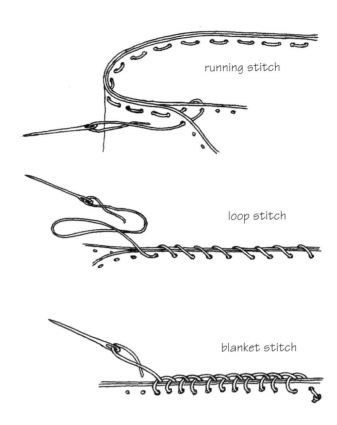

Basic sewing stitches.

projects where heavy cowhide is used. With softer leather wherein a fine thread might be apt to cut into the edges of the stitch holes, a leather lace might be more appropriate than thread, stitched through rows of pre-punched holes. Sinew was also commonly twisted into thread and used by primitive clothing makers.

Making Patterns

Taking things apart is one of the best ways I know of to figure out how they are made. I like to use this methodology when assembling articles of clothing. An existing garment could be carefully disassembled and its pieces used to make a pattern for the new garment. It saves a lot of guesswork. In some situations, one might have to create a new pattern. Yards of sturdy paper sold as inexpensive drop cloths in the

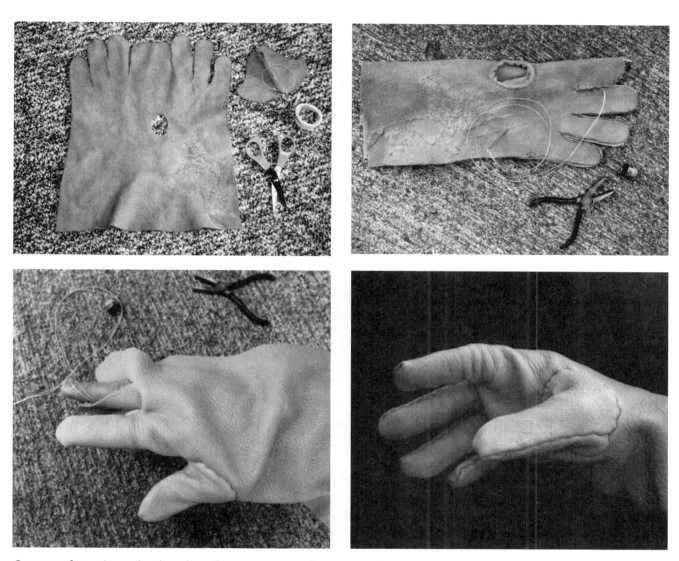

Sequence for making a leather glove. First, a one-piece fold-over pattern was drawn and cut out of elk hide, with a pilot hole cut for the thumb and a separate thumb piece. Next, the body and thumb piece were sewn up inside out, so that the stitches would be on the inside of the final product. It's good to test for fit periodically during the project, and adjust things as needed.

painting industry work especially well for making clothing patterns. Large paper grocery bags might also be opened up and used for patterns. The pattern pieces can be cut out of the paper (if not taken directly from disassembled used clothing), and stapled or clipped together to simulate sewn seams in order to test the paper article for fit, then disassembled and pinned to or set flat against the leather or cloth material and traced around with pencil or other marker to transfer the pattern.

The Sewing Kit

The essential tools of a basic multipurpose sewing kit include sewing or lacing needles, thread or lace, stitching awls, thimbles, clamps, clips or pins, scissors or a sharp knife, measuring tape or cord, markers, and small pliers. The pliers can be especially useful for pulling the end of a needle through tough leather.

Needles, awls, and thimbles can be handmade without too much difficulty.

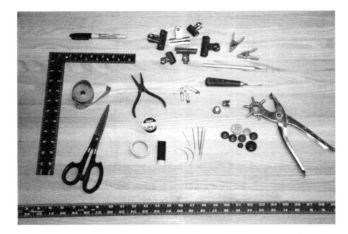

Some basic tools that will come in handy during the production of clothing.

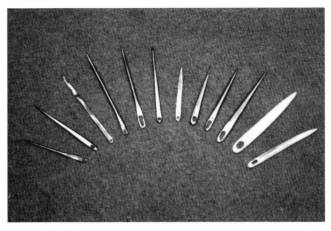

Makeshift sewing and lacing needles made from nails, saw blades, sheet brass, and bone.

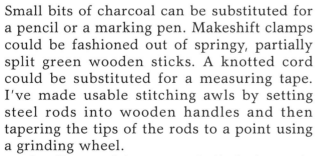

A small homemade sewing kit.

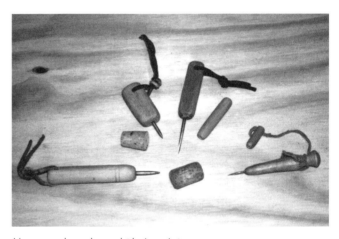

Homemade awls and their point covers.

Small bits of charcoal can be substituted for a pencil or a marking pen. Makeshift clamps could be fashioned out of springy, partially split green wooden sticks. A knotted cord could be substituted for a measuring tape. I've made usable stitching awls by setting steel rods into wooden handles and then tapering the tips of the rods to a point using a grinding wheel.

Small screwdrivers can similarly be made into awls simply by grinding the tips to a point. I've made sewing needles out of small nails, sheet metal, hacksaw blades, and even pieces of bone. Survivors would have to figure out ways to modify and use whatever they have, can find, or can make.

Material

Suitable clothing material can come from a number of different sources. Fabric from worn-out clothing can often be recycled. Shop rags, beach towels, tablecloths, curtains, drop cloths, and blankets could all be converted into clothing if needed. Blanket wool lends itself well as material for numerous projects, being heavy enough to offer excellent protection against the elements while also having generally good wear characteristics compared to many other common materials. In my experience, blanket wool is fairly easy to work with and sew together.

Cotton material is best washed in hot water to preshrink it before tracing the

Hooded poncho, mittens, gourd canteen cover, and two small bags all made from blanket wool.

A palm thimble made from a tough piece of cowhide.

patterns or cutting the pieces. Also, fabric for clothing looks better and lasts longer with hemmed edges. To prevent the woven threads from unraveling where the material has been cut, simply fold the last half-inch or so of all outer edges back and stitch them in place along the in-turned edges. This could be done before connecting the separate pieces or after the garment has been sewn together, depending on personal preference. Obviously, hems are not needed with animal skins, only with woven fabric.

Once we have created a pattern and obtained the fabric pieces of the correct size, the next order of business is getting the pieces connected by sewing them together.

The most important thing during the process is planning ahead. It can be frustrating to get far into a project, only to discover that an important step was omitted. Think about where you're going with each phase, and how it best fits into the sequence order. For example, it would be a shame to get two pieces sewn up before remembering that another piece was supposed to go between them. Having to remove stitches and undo a segment of work and start part of it over is not the end of the world, obviously, but it's a lot easier to think ahead and do it right the first time.

If you assemble articles of clothing inside out, the stitches and seams will be on the inside of the final product, which makes the clothing look more refined and keeps the stitches mostly hidden and less exposed. This may not be such an issue with clothing made from animal skins, which often tend to have a primitive look, anyway.

Sewing soft cloth is quite different than sewing leather. It is often preferable, and sometimes even necessary, to prepunch the stitch holes in leather. A sharp awl is a handy tool for the leather crafter, and I have found awls to be useful when sewing thick canvas as well. With my slow method I normally poke the holes as I go along, just one or two holes at a time ahead of the stitches. This ensures that the two separate pieces can be kept aligned during the sewing. When stitching leather or other heavy fabric, it's often useful to use a palm thimble—a scrap of very thick, tough cowhide or rawhide with a thumbhole. It fits into the palm of the needle hand and is used to help push the back of the needle through the tough leather. I normally like to keep a pair of pliers handy as well, for gripping the pointed end of the needle and pulling it on through. When thick pieces of leather are to be laced together, a leather hole punch that cuts larger holes can be a handy tool.

Before sewing separate pieces of material together to create a complete garment, it is

top piece

drawstring

Four pieces make the legs. Sew together and then turn inside out before attaching top piece.

ends to be hemmed

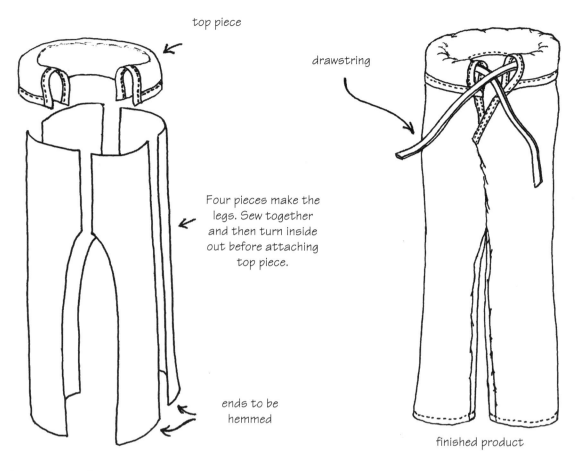

finished product

Simple drawstring pants.

normally very helpful to pin, clip, or clamp the seams together at intervals. This helps keep the seams properly aligned so that everything fits together correctly. Straight pins or safety pins can usually be used to hold fabric pieces together, while small clamps are more appropriate for leather products. Small spring clamps are sold in office supply stores, and I have found them to be especially useful with most of my leather and fabric projects.

SIMPLE CLOTHING PROJECTS

Modern clothing features a lot of zippers and snaps for closures; I prefer simple buttons and drawstrings wherever possible. Buttons are incredibly simple to manufacture out of wood, bone, antler, plastic, brass,

rawhide, or seashells. They are fairly reliable, perfectly functional in most applications, and much easier to replace or repair under basic conditions than are zippers. A simple button might be made by sawing a thin disk off the end of a hardwood dowel with the desired diameter, and then drilling two small stitch holes through it near the center. It can be finished by lightly sanding the edges.

Pants

Handmade pants can be closed at the fly and tightened around the waist using the drawstring system. Simply cap the top rim of the pants with an easily fashioned waistband tube having double thickness for strength, with a hemmed opening in front. A leather

A hooded poncho made out of a blanket. Simple to make, warm, and one size fits all.

thong or cord is then fed through this and used to tie the garment closed. Pants or trousers can be visualized as a bag with two tubes for the legs.

Poncho

One of the simplest garments I know of is a poncho. A poncho is really just a big sheet of some type of material, rectangular in shape, and folded in half with a hole in the center where it fits over the head. A very basic but functional poncho might be made out of a sheet of plastic, a blanket, or a small canvas tarp. In its most basic form, the poncho will offer some protection from wind and rain. It can be made more effective by adding a hood.

Shirt

A simple shirt can be made by starting with the poncho concept of folding over a piece of material with a hole in the center for the head to fit through. Tubes for the sleeves can be sewn on at the shoulders, and the trunk section can be sewn closed on both sides under the sleeves. The neck hole can be rimmed with a collar piece. There are certainly more conventional shirt designs, but this is easy enough and creates a functional garment.

Moccasins

Moccasins could be especially valuable to survivors after the supplies of more contemporary manufactured footgear are used up, because moccasins are more easily handmade and they are also very serviceable in a variety of climates and situations.

The variations in basic moccasin designs are surprisingly numerous. Certainly some styles are more complicated to make than others, and some might be more practical than others for certain applications, but any style of moccasin beats bare feet or worn-out shoes, in my view. Many designs should not be too difficult for the average person to build.

Bart and Robin Blankenship describe a practical method for making center-seam boot moccasins in their book, *Earth Knack*. They suggest making a custom-fitted pattern for the

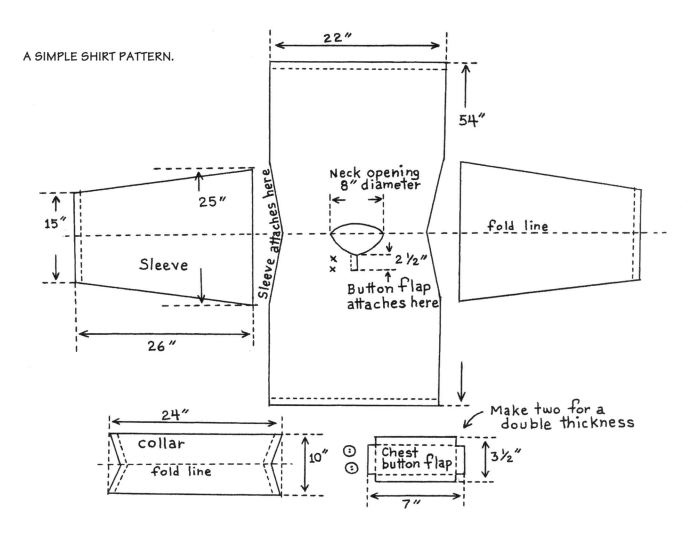

A SIMPLE SHIRT PATTERN.

22"

54"

25"

15"

Sleeve

26"

Sleeve attaches here

Neck opening 8" diameter

x x
x

2 ½"

Button flap attaches here

fold line

Make two for a double thickness

24"

collar

fold line

10"

Chest button flap

3 ½"

7"

Sturdy footwear, like this pair of three-piece moccasins, is a must, no matter in what conditions you're living.

A center-seam high-top moccasin.

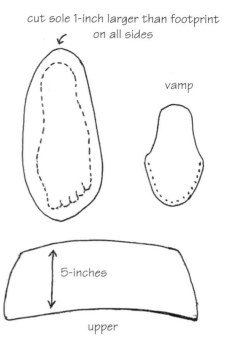

cut sole 1-inch larger than footprint on all sides

vamp

5-inches

upper

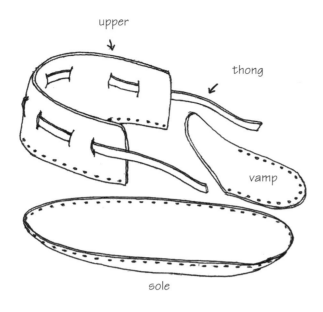

upper

thong

vamp

sole

Three-piece moccasin.

make a paper pattern to check size and fit

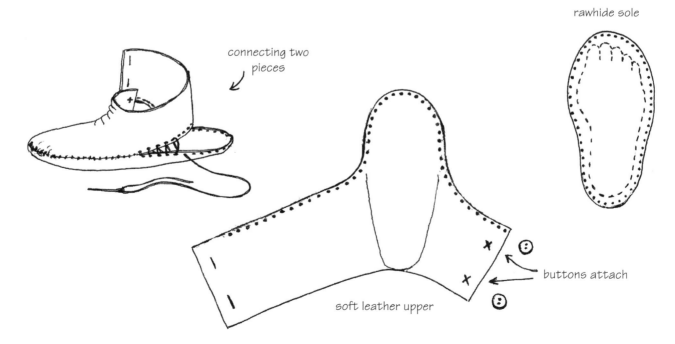

connecting two pieces

rawhide sole

buttons attach

soft leather upper

Two-piece moccasin.

boot by standing in stocking feet, and wrapping duct tape around the sock. The desired seams are then drawn over the tape with a marker and this molded cast is carefully cut off the foot with scissors along the seams to form a pattern that can be traced onto the chosen hide. The hide is cut where marked, and the seams are sewn together. A thicker sole can then be sewn to the bottom. I found that this design makes a very useful and comfortable boot moccasin.

A slightly simpler design is a basic three-piece moccasin, which was apparently popular with some of the Indians in the Southwest. It consists of a single sole of rawhide or heavy cowhide, a softer upper wrapping around the back, and the tongue piece, or vamp. A leather thong is used to tie the shoe securely onto the foot.

One of the most important aspects whenever assembling moccasins is properly puckering the toe. With most double-sole designs, the bottom sole can be replaced when it wears out, extending the life of the shoe. One should also factor in the stretchiness of animal hides when building moccasins. Typically, a moccasin will be designed either to wear with socks or without, and this is best considered before starting the project.

Moccasin pieces are sometimes sewn together with thread, and sometimes they are laced. If holes are to be prepunched for lacing, the trick will be to keep all of the corresponding holes aligned; an extra hole somewhere can throw everything off. My biggest challenge with the center-seam boot design was sewing on the outer sole the way I did, which involved having one of my hands always working inside the toe where I couldn't watch what I was doing. Perhaps the outer sole should be sewn on before sewing the seams closed on the upper section; however, replacing a worn-out bottom would require doing it the hard way again, unless the entire shoe was disassembled first.

In arctic regions, clothing was commonly made from animal furs. Assembling fur clothing with the fur on the inside provides better insulation and comfort than with the fur turned out, and also protects the soft fur from excessive wear.

ENDNOTE

1. A number of survival and primitive living books describe methods for tanning hides. *How to Stay Alive in the Woods* by Bradford Angier, *Outdoor Survival Skills* by Larry Dean Olsen, *Wilderness Living* by Gregory J. Davenport, *Tom Brown's Field Guide to Living with the Earth* by Tom Brown, Jr. with Brandt Morgan, *Live Off the Land in the City and Country* by Ragnar Benson, and *Earth Knack* by Bart and Robin Blankenship are some of the books I've run across that provide discussions on tanning skins. For a more thorough study of the subject, one might wish to read *The Complete Book of Tanning Skins and Furs* by James Churchill, *Leather Makin'* by Larry J. Wells, *Home Tanning and Leathercraft Simplified* by Kathy Kellogg, *Deerskins into Buckskins: How to Tan with Natural Materials* by Matt Richards, *Blue Mountain Buckskin: A Working Manual* by Jim Riggs, or *How to Tan Skins the Indian Way* by Evard H. Gibby.

CHAPTER 8

A Barter Economy

The utility value of hatchets and axes of all kinds cannot be denied.

Historically, whenever economic and social structures have broken down to the point where the public's confidence in its government has drastically deteriorated, the society's reliance on official currency has also deteriorated. People tend to find other mediums for trading, such as usable goods and services or the currency of more stable nations. This has occurred repeatedly in the past. Germany's money became almost worthless following WWI, and Confederate money became worthless after the American Civil War, to cite just two examples. This condition creates havoc with a nation's economy.

The U.S. dollar fluctuates in value as measured against the currency of other nations on a regular basis. Meanwhile, its buying power over goods and services domestically declines over time in the condition we call inflation. It is important to understand that the *only* thing giving our money any value at all is our confidence in the system. The public's confidence in the stability of its government allows the official currency to trade as money.

In circumstances where anarchy replaces the state of order we expect to see under functioning governments, traditional currency becomes unreliable. The people then search for more universally accepted trade mediums.

This is one example of how a barter economy might develop. Extreme taxation is another potential driving force behind a possible barter system, as private exchanges of trade goods are nearly impossible for any government to monitor and tax.

To fully understand the nature of a barter economy, it is helpful to examine the evolution of the world's monetary systems. It apparently didn't take ancient civilizations very long to realize the need for some standard of exchange. In ancient Rome, soldiers were paid with salt. Very primitive cultures have used seashells, whales' teeth, and even pebbles as money, among other things. Money is merely an accepted medium of exchange that represents an agreed value intended to make trading more efficient.

Bartering obviously constitutes a very awkward system of commerce. An individual who wants a knife, for example, might have some extra arrows he is willing to trade, but he faces the challenge of finding another individual who needs some arrows and is willing to trade a knife for them. The guy with the extra knife might not need arrows. Maybe he needs something else that the guy with the arrows doesn't happen to have. In a money-based economy, people are able to buy or sell more reliably.

Several thousand years ago, societies began basing the valuation of their accepted money on precious metals, usually gold and silver, to establish a recognized standard for trading. Some of the earliest coins were in fact minted from silver, gold, copper, and electrum (silver and gold alloy). Coins of precious metal, or money that could be exchanged for precious metal on demand, are called *commodity currency*. Money that has no intrinsic value, such as paper money not redeemable in gold or silver, or the copper-clad coins common today, are known as *fiat currency*.

Gold and silver coins were standard trading mediums for centuries throughout the civilized world until the 20th century, when complex world economics pressured growing industrial nations into abandoning money with comparatively static value. Fiat currency was pumped into rapidly expanding economies, backed only by public confidence in government. Fiat currency can fluctuate in value as the result of a number of different factors.

Our money in circulation today has no intrinsic value. Coins for circulation have not been minted of silver since 1964. Our paper bills are no longer silver or gold certificates. If our government were ever to lose the support of the people, there would be really no reason to continue trading in U.S. currency. It is my belief that our money-based economy would be replaced with a barter economy. Other value standards would be sought for trading.

In situations of emergency survival where firearms might be depended upon for defense or hunting, needed gun maintenance tools, spare parts, and ammunition might all be in high demand. My list of the 12 most universally popular firearm cartridges is as follows:

.22 Long Rifle
9mm Luger (9x19mm Parabellum)
.223 Remington (5.56mm NATO)
.308 Winchester (7.62mm NATO)
.30-06 Springfield
.30-30 Winchester (30 W.C.F.)

12-gauge shotgun
.45 ACP (45 Auto)
.357 Magnum
.44 Remington Magnum
.22 Winchester Magnum (.22 WMR)
7.62x39mm Russian

There are plenty of other very popular handgun and rifle cartridges in common use, such as 40 S&W, .45 Colt, .270 Win, .300 Win Mag, 7mm Rem Mag, and scores of others, but my list of 12 merely includes the *first* cartridges I would stock up on for barter, given their universal popularity. It shouldn't be too difficult to find people who will trade for any of them.

I believe the key phrase pertinent to a barter economy should be "universal desirability." If you happen to be in possession of something that is universally desired, for whatever reason, then you would have effective bargaining chips. You should theoretically be able to trade for at least some of what you might want or need.

Choosing universally desired goods for trading may present the biggest economic challenge for survivors in a barter system. Different people have different circumstances, lifestyles, priorities, and needs. Whiskey, for example, might trade especially well in certain isolated regions where loggers, trappers, or miners may like to indulge and party when they get together, but it may not trade so well in an Amish community. Likewise, wristwatches and clocks might be important to city dwellers, whose lives and routines are wrapped around time schedules, but farmers and woodsmen operate closer to natural time cycles measured by the rising and setting of the sun and by the seasons. To many of them, mechanical timepieces might be seen as unnecessary luxuries.

Perhaps there really is no perfect "one size fits all" trade commodity, but certain things will no doubt come closer to meeting the purpose than others. You could arrange your list of possible barter goods by category. "Essentials" would include things that the average human being will at some point need. All people must eat to survive. Hence, certain storable food products might be listed under the essentials column. Things like soap, toilet paper, salt, and other needed consumables might be universally essential.

The next category would be for those kinds of things that might be desired by only certain people, but would be extremely precious to those people. Not everyone is a smoker, for example, but to a smoker, cigarettes are extremely important. I learned during field exercises in the Army that after

several days in the field, smokers will trade anything they have for cigarettes. I would categorize these as "precious nonessentials." Whiskey, wine, cigarettes, tea, coffee, and maybe things like chocolate could be listed under precious nonessentials. These might be considered luxury items to some, but nevertheless very valuable to others, especially during hard times.

As explained in previous chapters, I consider things like tools and hardware to be universally essential to normal human existence. My list would have a third category called "nonconsumable durables." In this group one might list miscellaneous tools, hardware, traps, and weapons. As these products would be more or less nonconsumable, with demand for resupply less frequent than that for items in the first two categories. For example, a shovel might eventually break, or a firearm might eventually get loose or rust and fail to properly function, but chances are these types of things won't be purchased as regularly as might be bags of dried beans or bars of soap. This is something to consider when planning stores for a barter economy.

This issue, frequency of resupply needs, might be weighed against the typically limited storage life of many consumables. A steel trap, hand tool, or firearm can remain in a dry storage area for many years without showing any signs of deterioration. While sacks of rice, nuts, dried beans, smoked jerky, and canned goods might store reasonably well under ideal conditions, time works against any kind of food products. Often after only six months or a year, many foodstuffs will have lost some of their original nutritional value and, depending on the type of packaging and preservatives used, might be subject to spoilage. This is the nature of consumable products.

The shelf life of consumables will surely be a major concern for traders. Conventionally packaged foods, drugs, and medicines normally include an expiration date. The simple fact that certain commodities don't have a very long storage life is enough, in my view, to severely limit their trade potential in a barter economy.

A starving person may be willing to trade something of great value for a loaf of bread, but the bread must be traded or consumed quickly before it is destroyed by mold. The relatively short lifespan of a product such as bread makes it a poor trade investment in most circumstances, even as important as it might be to someone initially.

A can of beef stew on the other hand, which has a longer shelf life than a loaf of bread, would clearly be the better trade item of the two. The individual who trades for the can of stew has more options. He can consume the stew right away just as he would have done with the bread to satisfy his immediate hunger, and this purpose represents a particular value. Or, he can save it for later when he might be even hungrier than he is right now. It is inside a sealed can and will last longer than an unsealed loaf of bread. This option could represent an additional value. Finally, he may not wish to consume the stew at all, but instead decide to keep it until he finds someone else who will trade eagerly for it. These examples should illustrate the value of shelf life and durability of trade goods in a barter economy.

It seems logical that the ideal barter stock would be an assortment of commodities that we expect will have universal desirability in a barter economy. Long-lasting goods should naturally be expected to have more trade potential in the market than goods with very limited shelf lives, because they offer traders more possibilities, as we have seen. Other considerations might include portability and availability. Goods that are difficult to transport, or goods that might be common and fairly easy to obtain, should be less desirable and possibly less tradable than goods with the opposite qualities. The conditions of supply and demand would surely affect the values of goods and services in a barter economy just as they affect prices in a money-based economy.

Preparedness books that address barter scenarios typically provide lists of trade goods expected to be in demand after a collapse.

Some of this is speculative, because none of us really knows the future or exactly which set of unique circumstances we'll be facing.

Certain ideas, however, based in part on accounts from history's frequent tribulations and also on careful consideration of possible worst-case scenarios, might have merit. Some of the things that I've found typically listed in books include toothbrushes, toilet paper, soap, dental floss, matches, pocket calculators, salt, canned goods, miscellaneous tools and hardware, firewood, candles, medicines, fish hooks, firearm ammunition and hand-loading components like primers, gunpowder, and bullets. I would also add disposable cigarette lighters, nail clippers, coffee, kerosene, batteries and flashlight bulbs, backpacks, and clothing. Knives, axes, razors, saw blades, scissors, sewing needles and thread, and basic raw materials might be included under the general category of tools and hardware.

Gasoline might be something to consider if you had a safe way to store it and if fuel preservatives could be added as needed. However, society may eventually abandon the gasoline engine in favor of new technologies and power systems that won't be oil dependent.

Knives and scissors are nonperishable products that would normally be considered nonconsumable as well, suggesting there may not be a huge demand for frequent resupply. One might think there exists a proliferation of these products in society already, which would keep the market oversaturated with supply for some time.

However, the essential nature of these tools has me speculating that there might always be a fairly strong and regular demand for them. Edges wear down from repeated sharpening over time. Blades occasionally break. Tools are sometimes lost. With fewer factories in operation during a post-collapse era, the demand overshadowing supply factor should eventually come into play. It is difficult for me to imagine living without certain basic tools such as knives and scissors.

Considering how easy it is right now to find quality used tools (and other useful

Right now, kitchen matches and .22-caliber rimfire ammunition are fairly inexpensive. These are the kinds of things some expect to trade as money after our world crashes.

Commonly needed items like toothbrushes, candles, fishhooks, fingernail clippers, and cigarette lighters could become precious after the supplies dry up.

Just about everyone on the planet will eventually find scissors useful.

We can expect that people will always value certain basic items like toilet paper or salt.

things) at garage sales at near giveaway prices, stockpiling now for barter stores in the future seems like a pretty good idea to me. Even if a Dark Age never comes, extra tools and supplies could always be traded or sold to others in times of poverty or prosperity.

Animals were commonly traded in barter economies throughout human history, and still are to some degree in some parts of the world. Horses, cattle, pigs, and poultry might all become more important to most of us in hard times, just as they currently are within the world's farming and rural communities.

Likewise, seeds and other agricultural products may also be in high demand. The main disadvantages animals and livestock have in any economy are the feed and care they require and the land space and skill necessary in raising them. Even so, leather, beef, milk, bacon, and eggs have always been important commodities.

Certain professional services, including skilled and unskilled labor, should also have value in a barter economy, just as they do in a cash economy. As noted earlier in this book, certain skills might be in high demand after a collapse. Doctors, dentists, mechanics,

Eventually, some standard medium of exchange would be sought. The values of gold and silver coins are universally recognized, they are highly portable, and can be stored indefinitely.

gunsmiths, engineers, toolmakers, electricians, and security guards might all suddenly become especially important, finding their services easy to trade in exchange for whatever they might need.

Gold and silver coins might actually be the ultimate trading mediums. They are highly portable, stable, and durable, and represent conveniently standardized measured currency. They possess their own intrinsic precious metals value, which is recognized around the world. If there exists a trade medium that has historically been more reliable, constant, or universal than the gold coin, I am certainly not aware of it.

Gold is uniquely shiny, attractive, and virtually impervious to corrosion. It is a rare element that has a number of industrial uses and is costly to mine. It is a highly malleable metal—an ounce of gold can be stretched into a wire more than a mile long. Gold has filled cavities in teeth, wired electronic circuits, decorated human bodies in jewelry, plated expensive kitchenware, been used in artistic inlays, and served numerous industrial purposes in addition to being minted into coins. I believe it is very likely that the world's demand for gold will always exceed the supply.

Silver, while not as precious as gold, nevertheless has certain desirable properties and has been valued by civilizations for literally thousands of years. It is likely that both gold and silver coins will always have fairly reliable values.

Whenever trading high-value merchandise, firearms, ammunition, or even food wherever people are desperately hungry, security could become a major issue. It might be wise to trade only with familiar people—preferably other traders with whom a long business relationship has been established—whenever possible. It might also be a good idea to store the trade goods separately from any supplies kept for personal use and to avoid giving other traders access to an entire inventory of merchandise. One should naturally always be cautious and mindful of potential security risks. Being armed is almost always preferable to being unarmed while dealing with others. The most secure system for storage of barter goods might involve a network of secret caches.

As we can see, a preparation strategy anticipating economic chaos could be as easy and affordable as bargain hunting at yard sales, or buying just a few extra supplies regularly during routine grocery shopping and saving the extra supplies for barter in bad times. It doesn't have to cost much to begin a store of trade goods if you start small and build the supply gradually over a period of time. It could be thought of as a type of savings plan.

Likewise, small gold pieces and pre-1965 silver coins might be affordable right now in small quantities, as an investment in security and really a hedge against hard times. Hyperinflation or economic chaos could surely put gold and silver way out of our reach later on.

My guess is that rare antiques and collectibles would likely become less desirable during a period when desperate people are struggling to survive. Necessities, recognized gold and silver coins, and utilitarian goods will probably have the most universally accepted trade value in a barter economy, to secure the survivors' most essential needs first.

We may see the day when having things of value to trade could mean the difference between being able to barter for food and needed supplies, or simply going hungry and doing without certain things we might really need for our survival.

CHAPTER 9

Adapting to a New Social Order

When government institutions collapse, or when civilizations dissolve into total disorder, there are generally always governing systems of some form or another that assume power to bring new social order. The new regimes either arise directly from the social chaos within the region, just as Hitler came to power during Germany's chaotic decades following World War I, or in some cases the locals will find themselves ruled by a stronger occupying foreign power. Order is often eventually established to some degree by some type of government or organization that has the means to enforce martial law. Individual freedoms are typically surrendered in favor of a more regulated system that imposes new order.

Governments, we can be sure, would eventually arise from a world of anarchy and chaos. These would evolve, just as they have evolved throughout human history, because some people are natural leaders and others are followers.

The big question then becomes: What kind of society and government system would eventually develop to replace what we have right now? Of course, it is impossible for us to know the answer, given the dynamics of complex people with ever-changing needs and expectations and the unpredictability of circumstances.

However, we can see by looking at history and human nature that certain social conditions are fairly predictable, following certain regional or world events. We can expect, for example, that immediately following a major crisis within most metropolitan areas, looting of stores and incidents of random violence will occur.

Impoverished segments of a population that exist on the verge of emergency in the best of times will experience an overwhelming desperation very quickly in a major disaster. I am not aware of any major cities in the world that are completely free of poverty. The people living on the edge are usually pushed over the edge during a catastrophe, and drastic measures are then necessary to control the resulting chaos. After a worldwide catastrophe such as a nuclear warhead exchange between nations, the majority of the survivors may fall into that category of impoverished, traumatized citizens desperately struggling in their day-to-day survival. Their world will have been changed forever.

We might expect that harsh conditions would give birth to harsh forms of government. In some regions we might expect to see warlords or gangs ruling over the local populations, as they frequently have in many third-world countries around the globe. Wherever organized central governments develop, we may see authoritarian regimes and dictatorships ruling with tyrannical control, relying heavily on the use of armed force to gain and maintain their power. This concept of a Dark Age social order might be seen as a very sinister system when viewed through the eyes of today's American citizens.

People living in a Dark Age will have to adapt to their new world, whatever kind of world it might be, if they want to survive.

Social structures could be quite different from what we are currently accustomed to. Our ability to adapt might depend to some degree on our understanding of the new system. If civilizations were to consist primarily of regional clans or tribes compelled to compete for resources, we could then expect to see the kind of tribal warfare that was common in ancient times. In this scenario, most people would belong to small groups, and rules and customs would naturally differ dramatically from one geographical area to another.

If, on the other hand, larger organized governments develop, as eventually they probably would, the people could eventually find themselves living as servants to a tyrannical power. I believe it is unlikely that any system established to bring order after a period of anarchy would be respectful of the rights of individuals. I think the more likely scenario would feature some type of police state, with relative order maintained under martial law. The "civilized world" might resemble a large prison in many ways, where the common citizens might be forced by terror and intimidation to follow the strict rules imposed by their rulers. This has occurred many times throughout world history.

It should not be difficult to imagine how racial tensions might play out in a chaotic Dark Age period. My guess is that people would tend to join others with whom they have the most in common, to form groups that help them survive in the harsh and dangerous world. I believe that groups comprised of different races, of different social classes, and of different religions, or those speaking different languages, would generally tend to be more divided by these differences than ever.

There are many instances throughout the history of mankind where people of certain races, creeds, or cultural identities have been persecuted. Indeed, widespread violence and atrocities against certain ethnic groups have occurred in recent years—more than half a century after the defeat of the Nazis—in parts of Africa, the Middle East, and in Eastern Europe, despite international efforts to prevent such horrors. The dark side of human nature is a reality we can't seem to escape.

While preparing for a future Dark Age, I believe it is worth carefully considering the social dynamics of humankind and the psychological effects of a system collapse. Many of the rules and customs of the current civilized world may not necessarily be suited to a very different post-collapse world, at least in the earliest stages. Any major disruption to the organization of an advanced civilization might take some time to restore, depending on the severity of the disruption, of course. But it certainly took the present world order literally thousands of years to evolve into what it is today. Like the old saying goes, Rome wasn't built in a day. (Let us also make note that the great Roman Empire eventually collapsed.)

If we take a look at the devastation caused by the Indonesian tsunami of 2004, we might get some sense of how a large percentage of a survivor population would likely fare in the short term following a catastrophic global event. As of this writing, almost two years later, many of the regions hit by the tsunami have not yet rebuilt or noticeably recovered from that natural disaster. And wherever available resources are severely limited after some major catastrophe, we can expect the kind of living conditions that exist in the tsunami-hit areas. Inadequate nutrition and sanitation, depressed economies, extremely limited public services and security, and the rampant spreading of diseases are all to be expected following a catastrophic disaster that interrupts an orderly system.

There is a certain degree of predictability to some of these things. Societies, governments, people, nature, and the world are all constantly experiencing changes. If we can count on anything at all, we can count on the fact that changes will occur, in everything we know. The world will not always exist the way we know it today. Forces of nature periodically reshape the landscape. Living things are always dying while others are being born. Life is constantly adapting to environmental changes. Natural cycles are occurring all the

time. The presumption that the industrialized world is too advanced now to ever experience another Dark Age is unrealistic in my view. The biggest questions might simply be when, to what severity, and exactly how a major upset to present norms will come about.

We can see that it is possible to prepare for a system collapse to some degree, even without having specific and detailed information, as has been contemplated throughout this book. But while we are stockpiling supplies and caching survival gear, educating ourselves, and improving our survival skills, we might also begin preparing ourselves for the psychological aspects of the sort of traumatic circumstances we are talking about.

We can see how religious faith has helped people throughout recorded history endure every kind of extreme hardship, psychologically preparing believers for the worst the world can throw at them. The forecasting of apocalyptic events on Earth is indeed a component of some religious doctrines. We often hear from those who study these things that people with strong spiritual convictions generally tend to handle adversity better than those without. Their faith seems to transcend the physical realm.

I think it's also fair to say that, generally speaking, the more self-sufficient the people are, the less affected they will be by social or economic chaos or other long-term calamities, and the harder it will be for dictators or warlords to rule over them.

Inevitably, people will find different ways for coping with the future, both physically and mentally, just as they have in the past. How survivors will adapt to the extraordinary living conditions within the altered world they encounter following a global cataclysmic event should be interesting. Some of us might even be around to experience it firsthand, if we're lucky (or unlucky, depending on the perspective). But if you suddenly found yourself immersed in a global Dark Age characterized by social chaos, severe economic depression, famine, deadly epidemics, and routine violence, what would your survival strategies be?

About the Author

As far back as Jim Ballou can remember, his interests have included survival activities—hunting, fishing, exploring, treasure hunting, backpacking, shooting, making and using tools, and just plain improvising. He is the author of dozens of magazine articles on various subjects.

Ballou spent four years in the U.S. Army as a paratrooper with the 82nd Airborne Division in the 1980s. During that time he had the opportunity to attend an interesting three-week course at the Jungle Operations Training Center in Panama.

For close to 30 years he has lived in northern Idaho, where his home library is a good collection of survival and back-to-basics-type books, along with his all-time favorite book, Daniel Defoe's *Robinson Crusoe*. He is a long-time blacksmith and enjoys shooting and collecting guns of all kinds.